T0305471

Network Function Virtualization

Network Function Virtualization

Concepts and Applicability in 5G Networks

Ying Zhang

This edition first published 2018
© 2018 John Wiley & Sons, Inc.

All rights reserved. No part of this publication may be reproduced, stored in a retrieval system, or transmitted, in any form or by any means, electronic, mechanical, photocopying, recording or otherwise, except as permitted by law. Advice on how to obtain permission to reuse material from this title is available at http://www.wiley.com/go/permissions.

The right of Ying Zhang to be identified as the author of this work has been asserted in accordance with law.

Registered Offices
John Wiley & Sons, Inc., 111 River Street, Hoboken, NJ 07030, USA

Editorial Office
111 River Street, Hoboken, NJ 07030, USA

For details of our global editorial offices, customer services, and more information about Wiley products visit us at www.wiley.com.

Wiley also publishes its books in a variety of electronic formats and by print-on-demand. Some content that appears in standard print versions of this book may not be available in other formats.

Limit of Liability/Disclaimer of Warranty
The publisher and the authors make no representations or warranties with respect to the accuracy or completeness of the contents of this work and specifically disclaim all warranties; including without limitation any implied warranties of fitness for a particular purpose. This work is sold with the understanding that the publisher is not engaged in rendering professional services. The advice and strategies contained herein may not be suitable for every situation. In view of on-going research, equipment modifications, changes in governmental regulations, and the constant flow of information relating to the use of experimental reagents, equipment, and devices, the reader is urged to review and evaluate the information provided in the package insert or instructions for each chemical, piece of equipment, reagent, or device for, among other things, any changes in the instructions or indication of usage and for added warnings and precautions. The fact that an organization or website is referred to in this work as a citation and/or potential source of further information does not mean that the author or the publisher endorses the information the organization or website may provide or recommendations it may make. Further, readers should be aware that websites listed in this work may have changed or disappeared between when this works was written and when it is read. No warranty may be created or extended by any promotional statements for this work. Neither the publisher nor the author shall be liable for any damages arising here from.

Library of Congress Cataloging-in-Publication Data applied for

ISBN: 9781119390602

Cover design by Wiley
Cover image: © oxign/Gettyimages

Set in 10/12pt WarnockPro by SPi Global, Chennai, India

Printed in the United States of America

10 9 8 7 6 5 4 3 2 1

Contents

List of Figures

List of Tables

Preface

Because of the emergence of user-friendly smartphones and the advances in cellular data network technologies, the volume of data traffic carried by cellular networks has been experiencing a phenomenal rise. One large cellular operator has reported a growth of 8000% of cellular data traffic over the past 4 years [1], and it grows to 10.8 exabytes/month by 2016, an 18-fold increase over 2011.

To meet the constantly increasing traffic demands while maintaining or improving average revenue per user (ARPU), operators are always seeking new ways to reduce their operational expenditure (OPEX) and capital expenditure (CAPEX). One way to improve the cost-effectiveness of infrastructure is to introduce the concepts of scalable infrastructure and elastic capacity on demand, using *virtualization* and *cloud computing*, which have become the cornerstones of many successful large-scale IT infrastructure management.

Thanks to their successful deployment in Internet companies such as Google and Amazon, emerging cloud computing technologies have started to be borrowed and deployed in the telecom network, in the form of *software-defined networking* (SDN) and *network functions virtualization* (NFV). The SDN design introduces a separation between the control and forwarding components of the network. Among the use cases of such architecture are the access/aggregation domain of carrier-grade networks, mobile backhaul, cloud computing, and multilayer (e.g., Internet Protocol (IP), Ethernet, Optical Transmission Network (OTN), and wavelength division multiplexing (WDM)) support, all of which are among the main building blocks of today's network infrastructure. Unlike the traditional network architecture, which integrates both forwarding (data) and control planes on the same box, SDN decouples these two and runs the control plane on servers that might be in different physical locations from the forwarding elements (switches). It allows making the forwarding platform simple and bringing network's intelligence into a number of controllers that oversee the switches. Tight coupling of forwarding and control planes in the traditional architecture usually results in overly complicated control plane and complex network management and is known to create a large burden and high bearer to new protocols and technology developments. Despite the rapid improvements in line speeds, port densities, and performance, the network control plane mechanisms have advanced at a much slower pace. Thus, with the adoption of SDN, the network becomes more programmable and agile. While SDN covers the revolution of routers and switches in the network, NFV handles other boxes in the network that performs more complex packet processing. NFV calls for the virtualization of network functions (NFs) currently provided by legacy middleboxes and gateways. Using virtualization and cloud technologies, NFV allows legacy NFs offered by specialized equipment to run in software on generic hardware. It makes it possible to deploy virtualized network functions in high-performance commodity servers of operator's data center, with the great flexibility to spin on/off the services on demand.

Thanks to their promising benefits in flexibility and cost savings, today, both the wireline and wireless cellular networks have shifted to *virtualization* and *cloud* technologies, from both research and industry point of view. With the fast-growing traffic demand, the next-generation cellular network, that is, 5G cellular communications system, has heavily replied on the cloud technologies, NFV, and SDN. The 5G cellular network is expected to be commercialized by 2020. These technologies have been expected to impact different parts of cellular networks ranging from radio access network (RAN) all the way to the core network.

The main motivation for offering this book stems from the observation that, at present there is no systematic source of information about the cloud technologies' usage in the cellular network, as well as the interplay of different technologies, the discussion of different design choices, and its impact on our future cellular network. In addition to providing the latest advances in this area, we also discuss several research topics that have been studied in the academic in these fields. By bringing the basic concepts and their practical deployment scenarios together, we believe that there is tremendous potential on facilitating a more complete understanding of the entire space.

List of Abbreviation and Acronyms

3GPP	Third-generation partnership project
ARPU	Average revenue per user
BRAS	Broadband remote access server
BS	Base station
BSS	Business support system
BTS	Base transceiver station
CAPEX	Capital Expenditure
CPU	Central processing unit
C-RAN	Cloud RAN
DC	Data center
DPI	Deep packet inspection
D-RoF	Digital radio over fiber
DSL	Digital subscriber line
eNB	evolved NodeB
eNodeB	E-UTRAN NodeB
EPC	Evolved packet core
ETSI	European Telecommunications Standards Institute
eUTRAN	evolved UTRAN
GGSN	Gateway GPRS support node
GPRS	General packet radio service
GSM	Global system for mobile communications
GTP	GPRS tunneling protocol
HSPA	High-speed packet access
HTTP	Hypertext transfer protocol
IaaS	Infrastructure as a Service
IDS	Intrusion detection systems
IMS	IP multimedia subsystem
InP	Infrastructure provider

IoT	Internet of things
I/O	Input/Output
IP	Internet protocol
ISP	Internet service provider
IT	Information technology
L1	Layer 1
L1VPN	Layer 1 VPN
L2TP	Layer Two tunneling protocol
L2VPN	Layer 2 VPN
L3VPN	Layer 3 VPN
LAN	Local area network
LTE	Long term evolution
LTE-A	LTE-Advanced
M2M	Machine-to-Machine
MAC	Media access control
MANO	Management and network orchestration
MIMO	Multiple input multiple output
MME	Mobility management entity
MSC	Mobile services switching centre
MWC	Mobile World Congress
NAT	Network address translation
NEP	Network equipment provider
NF	Network functions
NFV	Network functions virtualization
NFVI	Network functions virtualization infrastructure
NIC	Network interface card
NV	Network virtualization
O&M	Operations and maintenance
OPEX	Operational Expenditure
OSPF	Open shortest path first
OSS	Operational support system
OTN	Optical transmission network
P2P	Peer-to-Peer
PaaS	Platform-as-a-Service
PDCP	Packet data convergence protocol
PE	Provider edge
PGW	Packet data networks gateway
PON	Passive optical network
POP	point of presence
PSTN	Public switched telephone network

QoE	Quality of Experience
QoS	Quality of Service
ODL	OpenDayLight
RAN	Radio access networks
RNC	Radio network controller
RRC	Radio resource control
SaaS	Software-as-a-Service
SDN	Software defined networking
SGSN	Serving GPRS support node
SGW	Serving gateway
SLA	Service-level agreement
SONET	Synchronous optical networking
SP	Service provider
SSL	Secure sockets layer
TCO	Total cost of ownership
UE	User equipment
UTRAN	Universal terrestrial radio access
vEPC	virtualized EPC
VIM	Virtualized infrastructure management
vIMS	virtualized IMS
VLAN	Virtual area network
VLR	Visitor location register
VNF	virtual network functions
vNIC	virtualized NIC
VoIP	Voice over IP
VPN	Virtual private network
VSN	Virtual sharing network
WAN	Wide area network
WiFi	Wireless fidelity
XaaS	X-as-a-Service

1

Introduction

Internet has truly become the service of the digitized society now and delivers a broad range of services such as banking, e-commerce, social networking, media, content storage, and much more. Currently, there is a strong trend to penetrate coverage and usage of Internet by going mobile. The usage by individual is steadily increasing both in time of use and bandwidth demand of application in use. Still there is 80% of the global population that lacks access to Internet. Clearly, the technologies building Internet need to evolve in order to facilitate the steady growth by cost-efficient and sustainable means. Moreover, it is commonly recognized in the technical literature that the Internet has constraints in terms of mobility, quality of service, security, and scalability (e.g., due to IP address starvation and semantic overloading of IP addresses) even if patches exist for fixing any particular problem.

Over the past decades, the telecommunications industry has migrated from legacy telephony networks to telephony networks based on an IP network. This shift allows the mobile network operators to leverage the high bandwidth, multiplexing, innovative products, and services that have been long deployed and tested in IP network and then stimulates a new wave of revenue generation. Since the emergence of cellular data networks, the volume of data traffic carried by cellular networks has been growing continuously due to the innovation of mobile devices, mobile applications, the rapid increase in subscriber size, and cellular communication bandwidth. The trend of cellular data growth will continue to accelerate as technology and application availabilities further improve. Indeed, in 2009, mobile data traffic exceeded mobile voice traffic for the first time [2, 3], starting the new paradigm of mobile networks. The dominant usage of mobile network has shifted from low bandwidth voice and messaging traffic

Network Function Virtualization: Concepts and Applicability in 5G Networks, First Edition. Ying Zhang.
© 2018 John Wiley & Sons, Inc. Published 2018 by John Wiley & Sons, Inc.

to a diverse type of data traffic, ranging from web browsing, video streaming, and even online gaming. Existing report [4] has shown that the mobile data traffic is increasing at a rate of 60% per year. By the end of 2016, mobile traffic has surpassed the traffic generated in the wireline network [5, 6]. It is predicted that by the year of 2018, the yearly IP traffic globally will be 64 times of that in 2005.

The demand on mobile network infrastructure will be further increased as the market of Internet of things (IoT) grows. IoT devices and their applications that function without direct human intervention are rapidly becoming an integral part of our lives. IoT devices and applications have wide use cases in a variety of areas, including telehealth, shipping and logistics, utility and environmental monitoring, industrial automation, and asset tracking. It is predicted to be 25 billion devices – including fixed/mobile personal devices and IoT devices – connected to IP networks [5] by the year of 2020 with a seven trillion dollar market value. The Machine-to-Machine (M2M) communications will generate 80% more data than the human directly generated data in the year of 2018. The prosperity of IoT industry requires great support of the wide area wireless communication infrastructure, in particular, the cellular data networks.

To cope with the explosive cellular data volume growth and best serve their customers, cellular network operators need to design and manage cellular core network architectures accordingly. There are two main challenges that the cellular network faces. The first challenge is the increasing cost of network operation. While the most advanced cellular technologies today is the 4G technology such as long term evolution (LTE). However, the coexistence of multiple generations of cellular networks is unavoidable today. This is because the subscribers may gradually upgrade their devices over the years. Thus, the operators need to support multiple technologies over a long period of time and manage multiple networks simultaneously. The operational cost may not be covered by the revenue growth. The second challenge is the high demand of network capacity expansion. Many techniques have been proposed to expand the cellular network capacity such as multi-antenna technologies and Wi-Fi offloading architecture. Adding spectrum and deploying small/femto cells have also been used together to further expand the network capacity, which is, however, expensive and not easy to deploy. Moreover, the changes of infrastructure still cannot keep up with the exponential growth of traffic demand.

The current cellular infrastructure is not capable of addressing the explosive need of data demand. The main reason is that the resource is configured and allocated in a rather static manner. The resources are not utilized in an efficient way. The traffic demand, however, are highly dynamic, exhibiting a time-of-day phenomenon [5, 6]. Flash crowds happen often due to popular events. To fundamentally meet the traffic demand, blindly increasing the network capacity is not enough. We need to find ways to better utilize existing capacity.

The traditional business model of cellular carriers was based on revenues for telephony. The Internet was an over-the-top service, priced by online minutes or data volumes during the late 1990s. This has changed completely to a flat-rate-based model for Internet access, and operators deploy broadband networks to cover demands of the digital society, struggling to return revenues needed to deploy even higher speed networks. Governance and regulations have further limited profitability of operators. In order to increase revenues, operators are deploying a number of service-centric networks on top of the broadband infrastructure such as IPTV, demanding new functionality of carrier network. On the cost side, carrier operators would like to reduce the capital and operational cost significantly. In the following, we discuss the opportunities of leveraging existing cloud technologies for better resource utilization. We will review the related technologies and then outline the rest of this book.

1.1   Cloud-Enabled 5G: SDN and NFV

The Internet has successfully been growing for more than 20 years; the growth in demand has so far been met by introducing even larger and larger routers. This has been beneficial and to scale in public networks. However, in order to meet today's steadily growing demand for Internet access and other packet-based services, there is a present need to deploy more efficient packet networks also within metro and aggregation network domain. The attempt to copy the approach from the coarsely populated, but large core network sites and migrate to metro and aggregation network sites may not be the most cost optimal approach. It may be time now to split the router architecture in similar ways as was done in the traditional mobile core network, in order to penetrate the highly dense metro/aggregation networks.

Splitting the router control and forwarding plane forms the initial idea of software-defined networking (SDN).

The initial idea was born to decouple the routing intelligence software from simple forwarding hardware allowing, particularly for academic research networks and test beds, fast prototyping and evaluation of new control theories and algorithms [7]. It was part of the Clean Slate Internet Design initiative of Stanford University [8]. The target is to develop a system that is amenable to high-performance and low-cost implementations and capable of supporting a broad range of research, can isolate experimental traffic from production, and is consistent with vendors' need for closed platforms.

The key technical idea of SDN is to provide an open control interface to the operating system of the network device without compromising the details of the implementation, an important business aspect for equipment manufacturers. This is enabled by support of Open-Flow [8] in the operating system and is based on the Ternary Content Addressable Memory (TCAM)-based flow tables, most routers and switches make use of. In a classical router or switch, the fast packet forwarding data path and the high-level routing decisions in the control path occur on the same device. An OpenFlow-enabled switch separates these two functions. The data path portion still resides in the switch, while high-level routing decisions are moved to a flow controller, typically a standard server. The OpenFlow Switch and Controller communicate via the OpenFlow protocol, which defines operation and management (OAM) messages.

Besides this technical view, this split design will enable a cost reduction and new market opportunities by the basic principle of modularization. This is of high importance for supporting flexible network innovations because the development cycles of hardware and software components are extremely different, and the modularization supports a decoupling of the innovations from a market perspective. The right layering approach will enable high market volumes for specific modules (software or hardware).

The introduction of the SDN concept into real networks would have a profound impact on the way in which networks are built and operated. In order to understand and evaluate the practical implications of the general concept, it would be beneficial to first test it in research networks. Feedback from the experimental implementation will be crucial in improving the overall concept and allow taking the concept to further applications in networking. First trials are currently under way in selected US universities, which focus on the easy management and reconfiguration of research networks, for example, for applications in the field of Clean Slate research.

While SDN brings innovative evolution to network routers and switches, the network comprises other types of devices besides routers and switches. Network operators enforce network policies using a combination of switches and network functions (NF). Policies may be complex, such as ensuring that unauthorized users are prevented from accessing sensitive servers or malicious traffic is eliminated from the network. To do this, an operator could use a stateful firewall to ensure that only traffic initiated from within the network is permitted and in doing so protect users from malicious traffic. Indeed, today's networks heavily rely on a wide spectrum of NFs. The diversity and complexity of NFs have been further expanded as the proliferation of wireless devices and mobile applications. NFs offer a variety of valuable benefits, ranging from improving security (e.g., firewalls, intrusion detection systems, and deep packet inspection), improving performance (e.g., proxies, caches) and reducing bandwidth costs (e.g., WAN optimizers, video transcoder). However, despite their benefits, NFs come with high infrastructure and management costs. One important reason is their complex and specialized processing. As a direct result of this complexity, configuration errors are common – configuration errors comprise as much as 65% of the network outages [9]. Other reasons of their complexity come from the lack of standardized management tools across different devices and vendors. Moreover, there is a need to consider policy interactions between these appliances and other network infrastructure, which cannot be easily troubleshot.

To facilitate programmability and flexibility of NFs, in 2012, operators initiated a new concept, called network functions virtualization (NFV) within the European Telecommunications Standards Institute (ETSI) consortium [10]. Instead of building NFs in the form of proprietary hardware boxes, NFV calls for the virtualization of them. Using virtualization and cloud technologies, it allows legacy NFs to be flexibly deployed in the form of software on commodity servers. Sharing the same spirit of splitting the router's control plane from forwarding plane, the decoupling of NF software from the hardware facilitates a faster pace for innovations and shorter development cycles, and result in shorter time to market of new services.

1.1.1 Benefits

One important benefit of the SDN and NFV is the potential to open new business opportunities in network architecture, related systems, and hardware. Today, when a new network protocol or network application is invented, the entire industry needs to go through the tedious standardization process by established organizations such as the Internet Engineering Task Force (IETF), International Telecommunication Union (ITU), and Institute of Electrical and Electronics Engineers (IEEE), to name but a few. After the standardization process, the development cycle and extensive testing activities may last 18–24 months, ensuring the performance, reliability, and interoperability with other equipments. While the seamless interconnectivity plays an important role in the success of the Internet and its applications, however, the long duration of standardization process and development cycle has made today's networks difficult to change. This is particularly true for research and innovations related to test and validate novel networking approaches, for example, new or revised routing or forwarding schemes. Such modifications currently require the close collaboration between operators and system manufactures, and hence demand substantial resources in terms of time and manpower.

The essence behind SDN and NFV is the decoupling of functionality. Such decoupling offers the opportunity to change the current limitations. In SDN, modifications to the control plane can be independently done, while the switching hardware remain unaltered. In NFV, the separation of software and hardware of NFs allows each part to grow and innovate independently. The modularity design choice in SDN and NFV comes with a number of economic advantages, especially in the following areas:

- Lowering the entry barriers for control plane product and switching plane gears
- Independent resource/budget allocation of each module
- Providing high extensibility because each module has clear interface with the rest
- Enabling rapid prototyping, instead of waiting for every component to be ready
- Allowing new players to enter the market. There are mutual benefits to both traditional equipment vendors and network operators, enabling small players to grow

- Increasing the speed of innovation. One key obstacle to any network innovation deployment is the incremental deployability and incentives for the first movers. Seamless interoperability usually slows down the deployment process and decreases the incentives for first adopters. With the modular design, first movers can gain technological advantages more easily when deploying new networks and services.

1.1.2 Challenges

While SDN and NFV have great potentials as discussed earlier, applying them in production networks faces several key challenges.

- *Scalability*: The SDN and NFV architectures need to scale to support millions of subscribers, increasing traffic demand and future growth. It needs to be able to meet the necessity of adding more controllers, switches, and NFs to a network. There is a potential need for a hierarchy of control components, or a chain of NFs, and might take a peer-to-peer approach to further expand the scalability.
- *Performance*: Carrier network has high requirement on performance than enterprise or data center networks. The links are usually tens to hundreds of Gigabits per second capacity. The customers' perceived performance should meet the predefined service level agreement (SLA). In SDN, how to deal with the response time of the controller considering the need to configure flows and how to provide sufficient bandwidth using a combination of gears are important questions to be answered. In NFV, how to control and reduce the additional overhead introduced by traversing several NFs and how to hide the overhead of virtualization techniques are critical to meeting the performance requirement.
- *Openness*: One of the key questions to enable openness is at what level of abstraction should be defined for different interfaces. The trade-offs between modularity and extensibility need to be carefully examined. For example, how one can add additional features to interfaces without affecting other parts of the system.
- *Internetworking and interoperability*: On the one hand, the network will gradually change. Seamlessly interconnecting SDN/NFV networks with traditional legacy networks is critical to its deployment. The devices supporting the new technologies should be able to communicate with legacy devices using traditional protocols, and the performance should not be compromised in this process.

- *Sustainability*: Energy efficiency has drawn significant attentions in the carrier network recently, both for the sake of reducing cost and for meeting the governance regulation requirements. The ability to switch off functionality and hardware, which is currently not in use, and how to make most efficient use of available processing capabilities are the key enablers to an energy-efficient network.
- *Resiliency*: In evaluating a network design, the network resilience is an important factor, as a failure of a few milliseconds may easily result in terabyte data losses on high-speed links. In traditional networks, where both control and data packets are transmitted on the same link, the control and data information are equally affected when a failure happens. In the SDN and NFV deployment, we need to come up with strategies to handle individual components' failures and to perform failure recovery to meet the carrier grade's five nines availability requirement.

1.2 Supporting Technologies

In this section, we briefly highlight the high-level ideas of related technologies. We will go into details of each of them in the rest of this book.

1.2.1 Cloud Computing

Cloud computing is transforming the way people use computers and how application services are run. While it has been widely used as the platform of choice for many web services, it is also becoming the hosting platform of network services. Virtualization is the key underlying technology enabling cloud providers to host services for a large number of customers. With virtualization, the physical resources and the applications are loosely coupled. The cloud customer is able to dynamically provision infrastructure to meet the current demand by leasing resources from the cloud infrastructure provider. On the other hand, the cloud provider can leverage economies of scale to provide dynamic, on-demand infrastructure in the best cost-effective manner. Using virtualization techniques [11], multiple virtual machines from different customers share the same physical servers. This is called *multitenancy*, which allows independent customers to lease resources from the cloud provider. Virtualization initially started with host virtualization, such as virtual machines

(VMs) or container. Later, the same concept has been extended to network virtualization, network functions virtualization (NFV), and so on. These technologies move the networking industry from today's manual configuration to more automated and scalable solutions. They are complementary approaches that solve different subsets of network mobility problem. SDN and NVF are among other initiatives to move from the traditional cellular infrastructure toward a cloud-based infrastructure across multiple areas including radio access network, core network, backhaul, and operational/business support systems (OSS/BSS). We elaborate them more as follows.

1.2.2 Network Virtualization

Network virtualization gives each customer tenant its own network topology and control over the flow of its traffic. For sharing computing resources, people have been widely used the virtual machine as the standard abstraction. For network virtualization, however, the right abstraction is still a subject of ongoing debate. There are a diverse set of solutions, which differ in the level of concrete details they expose to the individual tenants. In the widely used Amazon EC2, the network is abstracted in the simplest form. All of a tenant's virtual machines are connected with each other using best effort network fabric. VMware [12] offers a simple "one big switch" abstraction, where different rules such as access control or traffic engineering can be programmed by the tenants. The rules are implemented at the network edge (in the virtual switches inside the host operating systems). As more applications move to the cloud, providers must go beyond simple sharing of network bandwidth to support a wider range of abstractions. Recently, SDN has become a natural platform for network virtualization, thanks to the standard interface between controller applications and switch forwarding tables. However, supporting a large number of tenants with different topologies and controller applications raises scalability challenges.

1.2.3 Network Functions Virtualization

Network Functions Virtualization is a recent trend prompted by technology availability that makes high-performance packet processing now possible on commodity systems reasonably well, and service providers' desire for virtualization of network functions such as routers, firewalls, network address translation (NAT), and so on to avoid vendor lock-in and ability to choose virtualized network functions from different vendors as per the requirements. NFV not only allows service providers the ability to compose network functions but also the elasticity to flex compute and storage resources required by different VNFs depending on the traffic patterns and distributions. It provides a new way to design, deploy, and manage network services. It decouples the network functions from purpose-built hardware, so they can run in software. Most of the current focus in this space has been on virtualizing more and more network functions and cloud-like orchestration systems for managing deployed VNFs.

1.2.4 Software-Defined Networking

SDN is a new approach to designing, building, and managing networks, which enables the separation of the network's *control plane* and *data plane*, which makes it easier to optimize each. SDN has the potential to make significant improvements to service request response times, security, and reliability.

In SDN, the controllers collect information from switches, and compute and distribute the appropriate forwarding decisions to switches. Controllers and switches use a protocol to communicate and exchange information. An example of such protocol is OpenFlow [13], which provides an open and standard method for a switch to communicate with a controller, and has drawn significant interests from both academic and industry.

In summary, NV, NFV, and SDN each provide a software-based approach to networking, in order to make networks more scalable and innovative. Hence, some common beliefs guide the development of each. For example, they each aim to move functionality to software, use general-purpose hardware in lieu of purpose-built hardware, and support more efficient network services. Nevertheless, note that SDN, NV, and NFV are independent, though mutually beneficial.

1.3 Outline of Chapters

This book is divided into five chapters and provides information on the different technologies that have been examined in the design of 5G next-generation networks.

In Chapter 1, we introduce the background of carrier networks, the challenges of cellular network operations, and the requirements for future network moving into cloud. We discuss the advantages of considering SDN and NFV in the telecom communication networks. We also outline the developing technology that are relevant to 5G cloudification design.

Chapter 2 reviews various wired network virtualization technologies and wireless virtualization. It then studies the state of the art in cloud computing, the service models, and its relationship with communication networks.

In Chapter 3, we provide a survey of the existing NFV technologies. We first present its motivation, use cases, and architecture. We then focus on its key use case, the service function chaining, and the techniques and algorithms.

In Chapter 4, we provide a review of the SDN technology and business drivers, describe the high-level SDN architecture and principles, and give three scenarios of its use cases in mobile access aggregation networks and the cloud networks. Furthermore, we provide discussions on the design implementation considerations of SDN in the mobile networks and the cloud, in comparison with traditional networks.

In Chapter 5, we review the case studies of NFV in the next-generation 5G network. In particular, we discuss several use cases of SDN and NFV in the packet core network and in customer premise equipment (mobile edge networks). In both case studies, we discuss the challenges and the opportunities.

The main objective of Chapter 6 is to set the stage of existing activities in the industry from both standardization and open source consortium point of view. The standards movement drives the future implementation and use cases of these technologies. Given telecom industry has traditionally taken the route of standardization followed by implementation, IETF, ETSI, and third generation partnership project (3GPP) play an important role in the future design of SDN and NFV. On the other hand, leveraging the experience of other application services, more and more network services are moving toward open source contribution approach to accelerate the functionality development cycle. Recently, there are open source communities formed in the network service area. In this chapter, we summarize their activities, relationships, limitations, and future directions.

2

Virtualization and Cloud Computing

2.1 Cloud Computing

Cloud computing has become a widely used computing model to support cost-effective and efficient data processing using commodity servers. Cloud computing makes effective use of distributed environments for tackling large-scale computation problems on vast data set. There are multiple challenges with cloud computing, such as virtualization, isolation, performance, scalability, privacy, and security. In this section, we first provide an overview of the architecture of cloud computing. Then, we will go deeper into various virtualization technologies and focus on the network virtualization.

2.1.1 Architecture

Public cloud providers use an on-demand, pay-as-you-go model of compute and storage infrastructure as well as platform services. Amazon Web Services (AWS) led the early cloud computing revolution, beginning with their S3 service in 2006. Their services have been adopted by companies large and small, from backups and archival storage in S3, to compute in EC2, Virtual Private Clouds, IAM authorization and authentication, and RDS managed databases, to name a few. For customers, these services are easy to add, easy to consume, and can lead to a sprawling, poorly documented infrastructure.

Cloud computing can be viewed as a layering architecture, as shown in Figure 2.1.

Network Function Virtualization: Concepts and Applicability in 5G Networks, First Edition. Ying Zhang.
© 2018 John Wiley & Sons, Inc. Published 2018 by John Wiley & Sons, Inc.

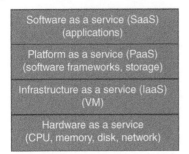

Figure 2.1 Cloud computing architecture.

- The hardware layer includes the physical resources in the cloud, that is, the hosting facilities, servers, switches, routers, hardware middleboxes, and power and cooling support. The *Hardware-as-a-Service (HaaS)* model means buying IT hardware or portions of data centers as a pay-as-you-go subscription service. Similar to other cloud computing layers, it shares the benefit of dynamically scaling up and down as the demand changes. The hardware is typically in the form of data centers, which consist of thousands of servers in racks. The HaaS provider needs to handle various hardware management issues such as configurations, fault tolerance, backup powers, and regular maintenance.
- The infrastructure layer is also known as the virtualization layer. The *Infrastructure-as-a-Service (IaaS)* provides computing resources as a service. Virtualization is an elegant and transparent way to enable time sharing and resource sharing on the common hardware. It allows customers to pay as they grow. By decoupling the hardware from the upper layer, it also helps make innovation faster and reduce the go-to-market time.
- The platform layer includes the operating system and application frameworks (e.g., Java framework) and other system components (e.g., data base and file system). Many popular cloud services operate at this level. For example, Microsoft Azure [14], Google AppEngine [15], and Amazon S3 [16] offer APIs for implementing typical web services.
- The *Software-as-a-Service (SaaS)* model means that the provider offers software on the common platform as well as the underlying database. Many traditional software companies (e.g., IBM, Microsoft, and Oracle) and the new players (e.g., Salesforce) are moving into this category. Cloud applications can automatically scale as the demand changes.

The layering architecture of cloud computing provides more modular design compared to traditional compute model. Resources are drawn whenever is needed on demand to fulfill a specific task. Unneeded resources can be relinquished, and the allocated resource is revoked after the job is done. Depending on the business model, cloud can be categorized to be *private cloud* where the data and processes are managed within the organization; *public cloud* where the resources and applications are provided/managed by a third-party off-site provider; and *hybrid cloud* where both internal and external cloud providers exist.

2.1.2 Types of Clouds

Originally synonymous with public clouds, today cloud computing breaks down into three primary forms: *public*, *private*, and *hybrid* clouds.[1] Each type has its own use cases and comes with its advantages and disadvantages.

Public cloud is the most recognizable form of cloud computing to many consumers. In a public cloud, resources are provided as a service in a virtualized environment, constructed using a pool of shared physical resources, and accessible over the Internet, typically on a pay-as-you-use model. These clouds are more suited to companies that need to test and develop application code and bring a service to market quickly, need incremental capacity, have less regulatory hurdles to overcome, are doing collaborative projects, or are looking to outsource part of their IT requirements. Despite their proliferation, a number of concerns have arisen about public clouds, including security, privacy, and interoperability. What is more, when internal computing resources are already available, exclusive use of public clouds means wasting prior investments. For these reasons, private and hybrid clouds have emerged, to make the environments secure and affordable.

Private clouds, can be defined in contrast to public clouds. While a public cloud provides services to multiple clients, using a shared infrastructure, a private cloud, as the name suggests, ring-fence the pool of resources, creating a distinct cloud platform that can be accessed only by a single organization. Hence, in a private cloud, services and infrastructure are maintained on a private network.

---

1 Some add a fourth type of cloud, called *community cloud* [17]. It refers to an infrastructure that is shared by multiple organizations and supports a specific community. The health-care industry is an example of an industry that is employing the community cloud concept.

Private clouds offer the highest level of security and control. On the other hand, they require the organization to purchase and maintain its own infrastructure and software, which reduces the cost efficiency. Besides, they require a high level of engagement from both management and IT departments to virtualize the business environment. Such a cloud is suited to businesses that have highly critical applications, must comply with strict regulations, or must conform to strict security and data privacy issues.

A hybrid cloud comprises both private and public cloud services. Hence, it is suited to companies that want the ability to move between them to get the best of both worlds. For example, an organization may run applications primarily on a private cloud but rely on a public cloud to accommodate spikes in usage. Likewise, an organization can maximize efficiency by employing public cloud services for nonsensitive operations while relying on a private cloud only when it is necessary. Meanwhile, they need to ensure that all platforms are seamlessly integrated. Hybrid clouds are particularly well suited for E-commerce since their sites must respond to fluctuating traffic on a daily and seasonal basis. On the downside, the organization has to keep track of multiple different security platforms and ensure that they can communicate with each other. Regardless of its drawbacks, the hybrid cloud appears to be the best option for many organizations.

In Table 2.1, we enlist the main benefits and risks associated with each type of clouds. Understandably, security is one of the main issues in cloud computing. There are many obstacles as well as opportunities for cloud computing. Availability and security are among the main concerns [19, 20].

2.1.3 Challenges

Being a disruptive technology, cloud computing has gained significant momentum in the past decade. However, it still faces several challenges with regard to performance, security, privacy, and interoperability. In the following, we discuss the challenges from these aspects:

- *Guaranteed performance*: Compared to the dedicated resource allocation in traditional compute model, cloud computing dynamically allocates resources on demand. While this is a key feature that enables multiplexing and elasticity, it also introduces serious concern on the application perceived performance. To provide

Table 2.1 Cloud computing: benefits and risks.

Cloud type	Benefits	Drawbacks
Public	• Low investment in the short run (pay-as-you-use) • Highly scalable • Quicker service to market	• Security: multitenancy and transfers over the Internet [18] • Privacy and reliability [18]
Private	• More control and reliability • Higher security • Higher performance	• Higher cost: heavy invest in hardware, administration and maintenance • Must comply with strict regulations
Hybrid	• Operational flexibility: can leverage both public and private cloud • Scalability: run bursty workloads on the public cloud • Cost-effective	• Security, privacy and integrity concerns

guaranteed performance, not only the compute resources (e.g., CPU and memory) should be allocated sufficiently, but also the networking resources (e.g., bandwidth and low latency) should be satisfied. Many existing cloud platform uses virtualization to provide isolated compute resources. However, the networking is still shared in a best effort manner across all the tenants, which can become a severe bottleneck when the network is congested. In addition, sharing the same physical server may also introduce additional delay due to context switching. The provider needs to have ways to monitor applications' performance so that the service-level agreement (SLA) can be met.

- *Security and privacy*: Data integrity and security is a big challenge for cloud platforms, especially for public cloud providers. Customers' data are competitive assets, and it is sensitive. Encrypting everything in the cloud will introduce additional overhead and may create inconvenience for security monitoring applications.
- *Fault tolerance*: Upon failure, the cloud platform needs to minimize the disruption to the applications and services. Most cloud platform uses seamless migration techniques such as VM migration to restart the application on another physical instance. However, certain infrastructure failures, for example, top-of-rack switch failure,

power outage, router failure, or even entire data center failure, are hard to bypass.

- *Resource management*: One of the most attractive features of cloud computing is the ability to acquire and release resources dynamically as the demand changes. The cloud provider needs automated resource management methods to effectively allocate and relinquish resources while minimizing the operational cost. To achieve that, the operators need tools to accurately monitor the SLAs with low overhead. Upon resource contention, they need to be able to sponge new resources and rebalance the demand across all available resources. Mapping the SLAs to low-level resources (CPUs and memory) is itself a challenging problem. Multiple types of resources may be dependent and thus a simple linear mapping may not be sufficient.

- *Interoperability*: Many companies have their own private cloud and would like to migrate to public cloud gradually. Some applications may need to use public cloud resources when the private cloud's resource is insufficient. Cloud applications may need to run on multiple cloud platforms simultaneously for geo-distribution and performance purposes. All these scenarios require cloud platforms and applications to be interoperable in order to support seamless migration.

2.2 Host Virtualization

Virtualization is a technique that abstracts out the low-level details of physical hardware and provides a simple, virtualized interface to the high-level applications. A virtual machine (VM) usually refers to a virtualized server. Virtualization is the key enabler of cloud computing. It provides the capability of sharing the server clusters as a pool of computing resources and the capability of dynamically mapping virtual resource to customers and applications. In this section, we review a few existing host virtualization techniques.

2.2.1 Overview

As early as 1960s, the concept of virtual machine has been introduced by IBM for providing concurrent access to the server. Each VM provides an abstract interface and an illusion of accessing the physical server directly and mutually exclusive to other users. This concept nicely enables time sharing and multiplexing on expensive physical resources and thus reduces the operational cost.

2.2.1.1 Benefits

The first important benefit brought by virtualization is the *isolation*. An application may contain bugs, which could interfere with other applications running on the same system. Most malwares operate like that to cause other applications' crash, compromise other applications, and steal sensitive information. Placing these applications in isolation with others can help improve the stability and security guarantee of the entire system. The second benefit of virtualization comes from *performance*. VMs provide exclusive access to the resources and thus have better performance guarantees than on a shared infrastructure. Traditional multiuser system provides sharing at the user level. In the user-level sharing, different applications compete on the computation resources, memory, network, and disk I/O.

2.2.1.2 Use Cases

There are multiple usage of VMs beyond cloud computing.

- *Using VMs as sandboxes for testing and security purposes*: VMs are useful to provide a secure and isolated environment for running untrusted applications. The isolation property provided by VM can prevent the malicious code from accessing underlying operating system resources or accessing other VMs' data and code. It is often used as sandbox to analyze the malware and provide signatures for future patching. It is also used for running suspicious or third-party less trustworthy applications.
- *Multiple operating environments*: On the same physical server, it can provide multiple operating systems simultaneously to support diverse application needs. One physical server can run both Linux and Windows Operating Sysytems (OSes) without interfering with each other.
- *Hiding hardware details*: Hardware is evolving and a hardware change may result in problems/crashes in the application software. The VM interface hides the hardware details and thus hardware upgrades can be transparent to applications. It can also provide virtualized hardware primitives such as virtual network interface cards (NICs), virtual either adapters, and switches.

- *Application migration*: On the one hand, it can be used to consolidate workload from the underutilized servers together to reduce energy cost. On the other hand, it can move applications to newer hardware servers with minimum disruption through live VM migration technique.

2.2.2 Virtualization Techniques

Virtualization is essentially a mapping and emulating process. It maps the virtual resource to native hardware resource and then uses the latter for computation. According to where the mapping occurs, we will first explain the widely used hardware-level virtualization and then cover other layers' virtualization as well.

2.2.2.1 Hardware-Level Virtualization

Figure 2.2 shows two widely used virtualization architectures. The most important component in this architecture is called virtual machine monitor (VMM). VMM uses software to emulate the virtualized processor, I/O, memory, storage, and so on. It can support multiple VM instances simultaneously. Each VM has its application and OS layer, just as a normal compute model. The OS passes the instructions to the VMM for execution. On the left side, the VMM is a software layer sitting on the bare metal hardware. So this architecture is also called as stand-alone VM implementation. In the right side of Figure 2.2, the VMM runs as an application on the host OS. It can utilize the host OS's functionality such as memory management, process scheduling, and hardware drivers. Both architectures have been widely used in many virtualization products.

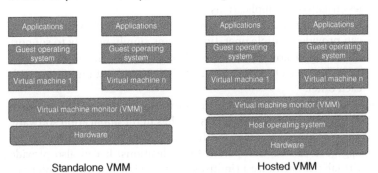

Figure 2.2 Virtual machine architectures.

The possibility of virtualization needs the support from the computer architecture level: the privileged instructions must trap so that the guest VM's privileged execution will be sent to VMM. The VMM will coordinate the scheduling of execution across VMs; either the instruction is executed on the native hardware or an emulated result is returned. VMM is in complete control of the entire system resource.

2.2.2.2 Other Virtualization Techniques

While the hardware virtualization is the fundamental building block to the cloud computing, there are other types of virtualizations. One of the well-known ones is the java virtual machine (JVM), which provides virtualization abstraction at the programming language level. As a platform-independent language, Java has a backend interpreter that performs platform-dependent translations. Programs run on the JVM platform are compiled to standardized portable binary format in the form of class files. The binary is executed by the JVM runtime, which emulates the JVM instructions.

2.2.3 Containers

Containers are a game changer in the virtualization space. Starting a new virtualization cannot get easier than this. Containers are easy to use, fast, and ensure real software portability. Each container has all its required runtime dependencies and configurations. Containers that run on the same engine share the same Linux kernel of the host server and are executed by Docker engine rather than hypervisor. Containers are much smaller in terms of size than virtual machine and have shorter start-up time. Therefore, Docker containers are considered as the agile and lightweight solution. While the lightweight OS solutions have existed for a few years, Docket container [21] is the one that leads to the mass adoption and the hype around containization. Docker container was initially designed as a Go language building runtime. Today, the Open Container Initiative builds the ecosystem around Docker container. Recently, a container standardization is discussed, and a runtime implementation is produced, called runC [22].

Figure 2.3 shows the architecture of container. Compared to the traditional VM architecture in Figure 2.2, the container does not have the per VM guest OS, which is usually tens of gigabytes. Each VM instance has a full guest OS image in addition to the binaries and libraries needed for the applications. It usually leads to high memory

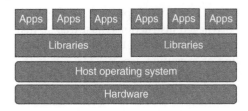

Figure 2.3 Docker container architectures.

and disk utilization and thus a slow start-up time. In contrast, in container, each application runs as an isolated process in user space on the host operating system, sharing kernel with other containers. Each application is usually on the order of megabytes, which is of much lighter weight. Therefore, container is a much more portable and efficient solution than VM. The resource isolation benefits are still provided by the host OS. Thus, container is a much lightweight virtualization concept.

The main technologies behind container are namespace, control group, and union file system (UFS). Each container runs a set of namespaces, which provides the isolation. Each container can only access its own namespaces. The control group is used to set up access control to hardware resources for each container. The UFS provides lightweight and fast layering, which is the building block of container. Docker container builds on top of these techniques into the container format in the form of libcontainer. In addition, Docker also supports traditional Linux container LXC.

The entire Docker stack in a clustered dockerized environment contains five major layers: Cluster Manager layer, Node (or host) layer in the cluster, Docker Demon layer that runs on each node, Containers layer created by the Docker Demon, and Applications layer that contains applications run on each container. Different Docker cluster managers are structured in different forms, for example, on Swarm, the containers are part of its structure, but on Kubernetes, there is only a Pod that represents a set of containers, and the user can control the Docker containers just through the cluster Pods.

2.3 Network Virtualization

Network virtualization refers to the technology that enables partitioning or aggregating a collection of network resources and presenting them to various users in a way that each user experiences an isolated and unique view of the physical network [23–25]. The abstraction of network resources may include fundamental resources (i.e., links and nodes) or derived resources (topologies) [23]. This technology may virtualize a network device (e.g., a router or NIC) a link (physical channel, data path, etc.), or a network.

With the advent of software-defined networking architectures and usage within virtualized cloud systems, the separation of virtual networking from the physical underlying resources pose new challenges. Most modern data center virtual network technologies use a tunneling system to emulate a network as an overlay onto a physical network infrastructure. Examples of these virtual networks include VXLAN and NVGRE. An end host's network traffic is admitted from the Virtual Tunnel End Point (VTEP) into the tunneling system (the virtual network). The tunneling system usually is agnostic to the endpoints. It usually uses a simple forwarding strategy in the physical network infrastructure, such as shortest path routing, to allow packets in the tunnel to reach their intended VTEP. The network policy and access control are usually performed at the virtual network layer. The mappings between the hosts and the VTEPs are maintained either in a centralized manner or distributedly shared across VTEPs. Such information can be maintained on edge switches or end host virtual switches, depending on different types of virtual networks. The routing and management of physical network is disjoint from the virtual networks. It is usually maintained by a different administrative domain. The physical network's configuration is usually more stable. This separation enables parallel management of different layers. Different policies can be applied without conflicts. However, because of the separation, the events, state changes, and faults on the physical network cannot be conveyed to the virtual network in real time, and vice versa. Network traffic on the virtual network is not provided with enhanced services that are typically available from the physical network. In addition, classic FCAPS (Fault/ Configuration/Accounting/Performance/Security) network management services are not interrelated between the virtual and physical domains.

For example, overlay tunnel end points that initiate very large flows onto the virtual network overlay have no awareness of the underlying network topology and cannot optimally route the data flow or provide quality of service (QoS) for differential treatment and, therefore, may cause congestion that could otherwise be avoided. A further example is troubleshooting in the event of a link failure on the physical network that affects an overlay tunnel carrying virtual network traffic. Without a federation between control systems, it becomes very difficult for administrators of the virtual network system to troubleshoot or detect the root cause of the issue.

In this section, we review different network virtualization technologies, discuss their pros and cons, and highlight the recent developments that are relevant to the SDN, NFV, and 5G.

2.3.1 Overlay Networks

An *overlay network*[2] is a logical network that runs independently on top a physical network (underlay). Overlay networks do not cause any changes to the underlying network. Peer-to-peer (P2P) networks, virtual private networks (VPNs), and voice over IP (VoIP) services such as Skype are examples of overlay networks [24–26]. Today, most overlay networks run on top of the public Internet, while the Internet itself began as an overlay running over the physical infrastructure of the public switched telephone network (PSTN). The Internet started by connecting a series of computers via the phone lines to share files and information between governmental offices and research agencies. Adding to the underlying voice-based telecommunications network, the Internet layer allowed data packets transmission across the public telephone system, without changing it.

P2P networks are an important class of overlay networks [27]; they use standard Internet protocols to prioritize data transmission between two or more remote computers in order to create direct connections to remote computers, for file sharing. P2P networks use the physical network's topology but outsource data prioritization and workload to software settings and memory allocation.

Although there are various implementations of overlays at different layers of the network stack, most of them have been implemented in the application layer on top of IP, and thus, they are restricted to the inherent limitations of the existing Internet.

2.3.2 Virtual Private Network

VPN provides private connectivity across different geographically separated sites. The sites can be offices of the same company that

---

2 Here, the network refers to a telecommunication or computer network.

spread across the country or around the globe. For these companies to expand their private network beyond their immediate geographic area, one needs a fast and reliable communication network among their offices. The traditional VPN is using the *leased lines* on top of a *wide area network* (WAN). A WAN could very well be the public Internet because of its reliability, performance, and security. A private WAN can be expensive because leased lines can be expensive. Moreover, leased lines are not viable for employees to access the corporate network wherever needed, for example, from home, from the road, or from other organizations.

A VPN is an assembly of two or more private networks or individual users that uses secured tunnels over a public telecommunication infrastructure, such as the Internet, for connection. A VPN is commonly used to provide distributed offices or individual users with secure access to their organization's network. VPN services enable remote access to the company Intranet at low cost, which enables the realization of mobile workforce. The VPN architecture supports a reliable authentication mechanism to provide easy access to the Intranet from anywhere anytime. The Intranet is accessible as long as the user has access to any edge media, including modems, ISDN, cable modems, DSL, fiber, Wi-Fi, and cellular wireless. The high-level architecture and process is shown in Figure 2.4.

There are multiple types of VPN services, depending on its implementation.

- *Link layer VPN*: Link layer technologies such as frame relay and Asynchronous Transfer Mode (ATM) can be used to implement

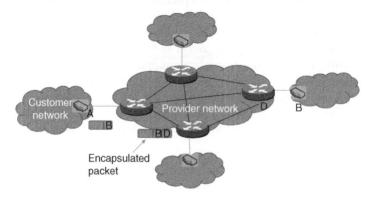

Figure 2.4 VPN illustration.

the VPN if the underlying network is based on the link layer technologies. The VPN is implemented using virtual circuits at the link layer. The frame relay frames or ATM cells are switched across nodes belonging to the virtual network. Virtual circuits are much cheaper than dedicated links and they are reconfigurable. The virtual circuits also have the advantage of providing SLAs.

- *MPLS VPN*: Multiprotocol Label Switching (MPLS) is a widely used protocol for VPN in the core network. Compared to the link layer VPN, MPLS VPN is much more scalable because they are based on connectionless architecture. Each customer site requires a customer edge router connected to a provider edge router in the provider network. MPLS also has advantages of providing performance QoS guarantees. Using Layer 3 routing protocols, for example, Border Gateway Protocol (BGP), the MPLS labels are distributed across all the customer and provider edge routers. Each IP packet is pushed with an MPLS label. The intermediate routers have a forwarding table that matches the label to a specific output forwarding port. The underlying infrastructure only performs the forwarding based on the label. The mapping between label and the original IP packet is maintained at the edge. MPLS is usually referred to Layer 2.5, between the link layer and the network layer.

- *IP VPN*: IP-based VPN is implemented at the network layer either by tunneling or by network layer encryption. A tunnel connects two points of a VPN across the shared network infrastructure. The network layer packets leaving a VPN node at one end of the tunnel are appended with an outer IP header with the destination address of the other end of the tunnel. The packets are then routed based on the outer IP address through the shared network infrastructure to the other end. The outer IP header is popped at the remote VPN node and the original packet is forwarded afterward. The tunneling method can encapsulate multiple protocol types on the same infrastructure. A disadvantage of the tunneling method is the management of a large number of tunnels. The network layer VPN can also use stronger secure encryption method, such as IPSec. IPSec can encrypt and encapsulate the original IP packet at the same time.

Based on the layer at which the VPN service provider's interchange VPN reachability information with customer sites, VPNs cam also be classified into three types: Layer 1 VPN (L1VPN), Layer 2 VPN

(L2VPN), and Layer 3 VPN (L3VPN) [23, 24, 28, 29]. While L1VPN technology is under development, the other two technologies are mature and have been widely deployed. Also, based on their networking requirements, enterprises can connect their corporate locations together in many different ways. These networking services can typically be viewed from three perspectives, which are demarcation point (or enterprise/service provider hand-off), the local loop (or access circuit), and the service core. Choosing Layer 2 or Layer 3 VPN will make a different impact on these three network service [30].

In an L2VPN, the service provider's network is virtualized as a Layer 2 switch, whereas it is virtualized as a Layer 3 router in an L3VPN [23]. In the former, the customer sites are responsible for building their own routing infrastructure. Put differently, in an L3VPN, the service provider participates in the customer's Layer 3 routing, while in an L2VPN it interconnects customer sites using Layer 2 technology.

As listed in Tables 2.2 and 2.3, both Layer 2 and Layer 3 services have their advantages and disadvantages. These are basically related to the differences of router and switch in computer networking; some of them are highlighted in Table 2.4.

2.3.3 Virtual Sharing Networks

VPNs and overlays are not the only types of virtual networks implemented so far; there exist other networks that do not fall into these two categories. Virtual *local area networks* (Virtual LANs) are

Table 2.2 Layer 2 VPNs: advantages and disadvantages.

Advantages

Highly flexible, granular, and scalable bandwidth

Transparent interface – no router hardware investment is required

Low latency – switched as opposed to routed

Ease of deployment – no configuration required for new sites

Enterprises have complete control over their own routing

Disadvantages

Layer 2 networks are susceptible to broadcast storms – due to no router hardware

No visibility from the service provider – monitoring services can be difficult

Extra administrative overhead of IP allocations – because of flat subnet

Table 2.3 Layer 3 VPNs: advantages and disadvantages.

Advantages

Extremely scalable for fast deployment

Readiness for voice and data convergence

"Any to any" connectivity – a shorter hop count between two local sites

Enterprises leverage the service provider's technical expertise for routing

Disadvantages

Increased costs – due to requiring customer router hardware

Class of service and quality of service usually incur additional fees

IP addressing modifications would have to be submitted to the service provider

Table 2.4 Router versus switch.

	Router	Switch
Definition	Connects two or more networks together and forwards packets between them	Connects many devices together on a network; more advanced than a hub
OSI layer	Network layer (L3) devices	Data link or network layer (L2 or L3)
Data form	Packet	Frame and packet
Address used for data transmission	IP address	MAC address
Table	Stores IP addresses in routing table	A network switch stores MAC addresses in a lookup table
Transmission type	At initial level broadcast; then unicast and multicast	First broadcast; then unicast and multicast
Function	Directs data in a network. Passes data between home computers, and between computers and the modem	Allow to connect multiple device and port can be managed; VLAN can create security also can apply
Speed	1–10 Mbps (wireless) 100 Mbps (wired)	10/100 Mbps, 1 Gbps

examples of these networks. While properly segmenting multiple network instances, such technologies commonly support sharing of physical resources among them. The term *virtual sharing networks* (VSNs) has recently been suggested for these types of networks [23].

Originally defined as a network of computers located within the same area, today LANs are identified by a single *broadcast domain* in which the information broadcasted by a user is received by every other user on that LAN while it is prevented from leaving the LAN using a router. The formation of broadcast domains in LANs depends on the physical connection of the devices in the network. Virtual LANs (VLANs) were developed to allow a network manager to logically segment a LAN into different broadcast domains. Thus, VLANs share a same physical LAN infrastructure, but they belong to different broadcast domains. Since it is a logical, rather than a physical, segmentation, it does not require the workstations to be physically located together. They can be on different floors of a building, or even in different buildings. Further, broadcast domain in a VLAN can be defined without using routers; instead, bridging software is used to define which workstations belong to the broadcast domain, and routers are only used to communicate between two VLANs.

The sharing and segmentation concept of the VLAN can be generalized to a broader set of networks, collectively called (VSNs). The key requirement for such networks is to share a physical infrastructure while being properly segmented [23]. For example, a large corporate may have different networks with specific permission for guests, employees, and administrators, yet all sharing the same access points, switches, router, and servers.

2.3.4 Switch-Based SDN Virtualization

Network virtualization enables multiple logical networks to share the same underlying physical network infrastructure in an isolated manner. Each logical network can have different addressing and forwarding schemes. The isolation and abstraction can be implemented at the switch level or the host level. First, we introduce the mechanism of implementing it at the network switch level using SDN. The advantage of this approach is that it does not rely on support from the end hosts. Once deployed, it can be used to support a large number of end hosts simultaneously. It also does not rely on specialized functionalities on the hardware, which then can be deployed on commodity switches.

There are multiple requirements to support network virtualization.

- *Autonomous management*: Each virtual network administrator has his own view of the topology and can freely define any routing scheme on top of it. The virtual network topology consists of nodes, links, and the connectivity between them. It can choose to use shortest path routing, multipath routing, or any customized routing mechanism. Inside the virtual network, any form of forwarding methods can be used, for example, Layer 2, Layer 3, MPLS, and Openflow.
- *Isolation*: For security and independence concern, the network visualization infrastructure should provide isolation to different customers. On the one hand, each virtual network should have separate routing entries. Failures or misconfiguration on one customer network may not affect other virtual networks.
- *Performance*: Each customer network can specify the amount of guaranteed bandwidth they want on each link. The underlying network should provide the capability of bandwidth division across multiple networks on each physical link.

One way to provide virtual network is at the switch level using SDN architecture. The seminal work in this category is Flowvisor [31]. Figure 2.5 shows the general architecture of this type of approaches. Each virtual network runs their own SDN controller, which can be called as user-level controller. The user controller has a northbound interface to accept the customer network's management input. It presents an abstracted topology to the user. Below the user controller, it employs an infrastructure-level controller that translates the user-level Openflow rules to the OF rules on the actual physical switches. To provide isolation, each switch can have multiple

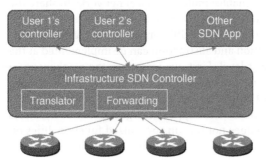

Figure 2.5 Switch-based SDN virtualization.

virtual tables to host the rules for different customer networks. This infrastructure controller essentially translates the virtual layer rules to the physical layer rules. Thus, it provides one level of abstraction, that is, the virtual layer is independent of the physical layer. If there is any change in the physical layer, the translation layer can hide the changes to the virtual layer. The user controller requires no modification and acts as if it was communicating directly with the physical switches. Since we will introduce SDN in more details in later sections, we keep the description here at a high level for now.

2.3.5 Host-Based Network Virtualization

Another way to implement network virtualization is using a host-oriented approach. Given the wide deployment and success of server virtualization techniques, this approach relies on the virtual switch in the hypervisor to control the forwarding behavior so as to provide an abstraction for virtual network. Among the work in this area, VMWare's NVP system [32] has been widely used in practice in a scalable manner. This approach is usually used in a multitenant data center, where a large set of physical hosts connected by a physical network under one administrative domain. Each host runs multiple VMs on the hypervisor, which has a virtual switch to handle the forwarding from and to these VMs. The notion of *network hypervisor* is then built using these distributed virtual switches. The network hypervisor provides an abstraction of control plane to the tenants. The tenants can use this control plane to define logical network elements such as logical switches and their forwarding rules. The sequence of logical elements forms a logical data plane. This logical data path is translated into rules and implemented in the software virtual switch. Figure 2.6 shows the architecture; there are essentially three segments in the network path: source hypervisor, tunneling, and destination hypervisor. There is a tunnel between any pair of hosts in the physical network; any typical tunneling protocol can be

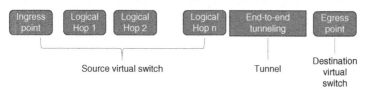

Figure 2.6 Host-based SDN virtualization.

used as long as the intermediate switches support it. For example, MPLS or GRE tunneling are good choices. Thus, the underlying network only needs to perform routing based on destination address, which is no different than ordinary IP networks. The logical data path is completely implemented in source virtual switch. The source virtual switch usually has multiple tables for the controller to install different logical hops. The destination VM decapsulates the packet and forwards it to the right VM according to the inner packet header.

While the design is simple, the main challenges for this design include the data path performance, the ease of management, and the scalability. Managing all pair tunneling itself is a challenging. Because the entire data path is implemented in the source virtual switch, the virtual switch's rule space management is also a key factor to the overall performance. Finally, providing a declarative management interface for the tenants' to flexibly specify their topology and policy is important for its deployment.

2.4 Wireless Virtualization

As a natural extension of wired network virtualization, wireless networks virtualization is motivated by the observed benefits of that in wired networks. However, while virtualization of wired networks and computing systems has become a trend, much less virtualization has occurred in infrastructure-based wireless networks [33]. Yet, the idea of virtualizing wireless access has recently attracted substantial attention in both academia and industry. It is one of the frontier research areas in computer science [23, 33, 34].

Wireless virtualization may refer to wireless access virtualization, wireless infrastructure virtualization, wireless network virtualization, or even mobile network virtualization [23, 34]. It is about the abstraction and sharing of wireless resources and wireless network devices among multiple users while keeping them isolated. Wireless resources may include low-level PHY resources (e.g., frequency, time, and space) or wireless equipment (e.g., a base station (BS)[3]), a network device (e.g., a router), a network, or a client hardware (e.g., wireless NIC).

---

3 A single physical BS can be abstracted to support multiple mobile operators and allow individual control of each RAN by having a separate vBS configured for each operator.

Thus, similar to wired network virtualization, wireless network virtualization software may reproduce logical channel and logical RAN (L1) in addition to logical switches and logical routers (L2–L3).

The motivations for virtualizing wireless networks are very similar, but not limited, to those of wired networks. First, as an extension of wired network virtualization, wireless virtualization can potentially enable separation of traffic to increase flexibility (e.g., in terms of QoS), improve security, and facilitate manageability of networks. Powerful network management mechanisms are particularly important in emerging heterogeneous networks. Second, it has a great potential to increase the utilization of wireless networks. This is important from both infrastructure and spectrum virtualization points of view. The former opens up the doors for the concept of IaaS so that one operator can use its own or other operator's underutilized equipment (e.g., BSs on the outskirts) in the congested sites, for example, in downtown. Spectrum virtualization can also provide better utilization; it may even bring more gain and is more valuable as spectrum is a scarce resource. Third, by decoupling the logical and physical infrastructures, wireless virtualization promotes mobile virtual network operators (MVNOs[4]). This allows decoupling operators from the cost of infrastructure ownership (capital and operation expenditures). Fourth, wireless virtualization provides easier migration to newer products and will likely support the emergence of new services. Last but not the least, it is a key enabler for cloud radio access network, which is expected to help operators reduce TCO and become greener.

Depending on the type of the resources being virtualized and the objective of virtualization, three different generic frameworks can be identified for wireless virtualization [34]:

1) **Flow-based virtualization** deals with the isolation, scheduling, management, and service differentiation between *traffic flows*, streams of data sharing a common signature. It is inspired by the flow-based SDN and network virtualization but in the realm of wireless networks and technologies. Thus, it requires wireless-specific functionalities such as the radio resource blocks scheduler to support QoS and SLA over the traffic flows.

---

4 MVNOs [35] are a new breed of wireless network operators who may not own the wireless infrastructure or spectrum, but give a virtual appearance of owning a wireless network. Basically, MVNOs resell the services of big operators, usually lower prices and with more flexible plans. Virgin Mobile is an example for MVNO.

2) **Protocol-based virtualization** allows to isolate, customize, and manage multiple wireless protocol stacks on a single radio hardware, which is not possible in flow-based virtualization. This means that MAC and PHY resources are being virtualized. Consequently, each tenant can have their own MAC and PHY configuration parameters while such a differentiation is not possible in a flow-based virtualization. The wireless NIC[5] virtualization [38, 39] where IEEE 802.11 is virtualized by means of the 802.11 wireless NIC, falls into this category.

3) **RF frontend and spectrum-based virtualization** is the deepest level of virtualization that focuses on the abstraction and dynamic allocation of the spectrum. Also, it decouples the RF frontend from the protocols and allows a single frontend to be used by multiple virtual nodes or a single user to use multiple virtual frontends. The spectrum allocation in the spectrum-based virtualization differs from that of the flow-based virtualization for its broader scope and potential to use noncontiguous bands as well as the spectrum allocated to different standards.

As noted, the depth of virtualization is different in these three frameworks and they are complementary to each other. From an implementation perspective, the flow-based virtualization is the most feasible approach with immediate benefits. It connects virtual resources and provides a more flexible and efficient traffic management. In all three cases, a flow-based virtualization is required to integrate the data. However, in the flow-based approach, the depth of virtualization is not sufficient for more advanced wireless communication techniques, such as the *coordinated multipoint* transmission and reception [34, 40, 41].

As a potential enabler for future radio access network, wireless virtualization is gaining increasing attention. However, virtualization of wireless networks, especially efficient spectrum Virtualization, is far

---

5 By means of a wireless NIC, which is basically a Wi-Fi card, a computer workstation can be configured to act as an 802.11 access point. As a result, 802.11 virtualization techniques can be applied to the 802.11 wireless NIC. Virtualization of WLAN, known as VirtualWiFi (previously MultiNet [36, 37]), is a relatively old technology. It abstracts a single WLAN card as multiple virtual WLAN cards, each to connect to a different wireless network. Therefore, it allows a user to simultaneously connect his machine to multiple wireless networks using a single WLAN card.

more complicated than that of a wired network. It faces some unique challenges that are not seen in wired networks and data centers. Virtualization of the wireless link is the biggest challenge in this domain [34, 42, 43]. Some other key issues in wireless virtualization are as follows:

- *Isolation*: Isolation is necessary to guarantee that each operator can make independent decision on their resources [44]. Also, since resources are shared in a virtualized environment, there must be effective techniques for ensuring that the resource usage of one user has little impact on others. In wired networks, this may only occur when every user is not provided with distinct resources, mainly due to resources insufficiency. *Overprovisioning* can solve the issue in such cases. It is not, however, a viable solution in wireless virtualization because *spectrum*, the key wireless resource, is scarce. To fulfill such requirements, sophisticated dynamic resource partitioning and sharing models are required.
- *Network management*: Wireless networks are composed of various radio access technologies (RATs), for example, 3G, 4G, and Wi-Fi. Similarly, a single wireless device is capable of accessing to multi-RAT. In such a multi-RAT environment, resource sharing is not straightforward. In contrast to network virtualization technologies that are mainly based on Ethernet, wireless virtualization must penetrate deeper into the MAC and PHY layers. Further, even in a single RAT environment, slicing and sharing is not easy because wireless channels are very dynamic in nature and an efficient slicing may require dynamic or cognitive spectrum sharing methods [45]. Hence, *dynamic network virtualization* algorithms must be considered.
- *Interference*: Wireless networks are highly prone to interference and their performance is limited by that. Interference is out there, particularly, in dense, urban area. This must be considered in slicing radio resources since it is not easy to isolate and disjoint subspaces. Especially, in a multi-RAT environment, if different spectrum bands of various RATs are shared and abstracted together, interference becomes even a bigger issue because interference between different RAT needs to be taken into account too. For example, a slice from Wi-Fi unlicensed spectrum could be assigned to an LTE user, causing unforeseen interference between LTE and Wi-Fi networks.

- *Latency*: [44] Current wireless standards impose very strict latency, in order to meet real-time application's requirement [46]. This mandates 5–15 ms round-trip latency in Layer 1 and Layer 2 of today's wireless standards and will be more stringent in the next generation (5G) [47].

There are also other concerns such as synchronization, jitter [48], and security [49].

2.5 Summary

Cloud computing has received significant momentum in the past decade and has been widely used in today's IT environment. Virtualization is the key technology that enables cloud computing. Host virtualizations such as VMs or containers have been extensively studied and widely deployed. From network virtualization's perspective, the traditional overlay network using encapsulation and tunneling is one type of virtual network. VPN is a form of popular virtual network for security applications. Recently, SDN has become another way to implement network virtualization. For example, one can define the virtual networks in the flow tables of the SDN switches. On the other hand, wireless virtualization has received a lot of momentum lately. The SDN approach brings the ability to do flexible traffic engineering and to gather more intelligence and statistics from the infrastructure.

3

Network Function Virtualization

Network function virtualization (NFV) represents a significant transformation for telecommunications/service provider networks, driven
by the goals of reducing cost, increasing flexibility, and providing per-
sonalized services [10]. It is rapidly emerging as the de facto approach
operators will use to deploy their networks. The promise of NFV will
play out over the next several years, and several challenges need to
be addressed to make that happen. Telcos have transitioned most of
their communications to standard IP networks and are now starting
to migrate most of their computing to industry standard servers. NFV
leverages on cloud computing principles to change the way NFs such
as gateways and middleboxes are offered. Different from today's tight
coupling between the NF software and dedicated hardware, the loosely
coupled software and hardware in NFV can reduce the upgrade cost
and increase the innovation flexibility. In addition, with the advent of
5G technology, there is a push toward developing novel real-time services in the areas of mobile payment, augmented reality, autonomous
driving, and Internet of things (IoT) [10].

Typical network functions (e.g., server load balancing (SLB), and
firewalling) in large-scale enterprise networks today are generally
provided on large vendor-proprietary devices. These are complex to
operate and manage, expensive to procure and maintain, and are
typically very underutilized. They also do not lend themselves to easy
automation since, because they are proprietary appliance based, their
application programming interfaces (APIs) are vendor-specific, making them difficult to integrate into a high-level non-vendor-specific
orchestration platform. Efforts to alleviate this complexity through
flexible network routing techniques such as SDN have led to only a
partial fix, as the physical switch devices controlled via SDN/OF lack

Network Function Virtualization: Concepts and Applicability in 5G Networks, First Edition. Ying Zhang.
© 2018 John Wiley & Sons, Inc. Published 2018 by John Wiley & Sons, Inc.

the intelligence to make decisions based on OSI Layers 5–7 and so are not able to deliver the higher level services provided by the load balancers and firewalls.

In NFV environments, monolithic complex network functions running on specialized hardware are decomposed into smaller functional units and dynamically orchestrated onto a virtualized cloud and edge infrastructure. These network functions include both control plane processing, such as telco equipments for signaling, authentication, and data plane processing, such as normal routing, access control, and quality of service (QoS). These network functions run as software in a virtual machine (VM), which is also called virtualized network function (VNF). Running as VMs among a large number of TOR switches in the data center allows these functions to scale on demand. It then helps reduce the operational cost for the telecom operators. It also has the potential to reduce the growth of appliance sprawl, particularly in multitenant data centers (DCs), providing power, space, and cooling savings for a more efficient and greener data center. A highly agile routing instructions that direct network packets through a sequence of VNFs is called service function chaining (SFC), which we will discuss more in later chapters. In this chapter, we focus on the basic building blocks of NFV.

3.1 NFV Architecture

To meet the increasing traffic demands while maintaining or improving average revenue per user (ARPU), network operators are constantly seeking new ways to reduce their OPEX and CAPEX. To this end, the concept of NFV was initiated within the European Telecommunications Standards Institute (ETSI) consortium [50]. NFV allows legacy NFs offered by specialized equipment to run in software on generic hardware. The key driving technologies behind that is cloud computing and virtualization. NFV makes it possible to deploy VNFs in high-performance commodity servers in an operator's data center, with great flexibility to spin on/off the VNFs on demand. The agility has brought NFV increased popularity by network operators because it can achieve high resource utilization. In addition, by decoupling the NF software from the hardware, NFV will facilitate a faster pace for innovations and result in shorter service development cycles. In summary, NFV has great potential in CAPEX/OPEX savings and ARPU increasing.

NFV covers a wide spectrum of middleboxes such as firewalls, deep packet inspection (DPI), intrusion detection system (IDS), network address translation (NAT), and wide area network (WAN) accelerators. It also covers a variety of network nodes such as broadband remote access servers data network gateways (S-GW/P-GW), mobility management entity (MME), home subscriber server (HSS), and virtual IP multimedia subsystem (vIMS) for virtual evolved packet core (vEPC). These are the critical devices in the mobile broadband and cellular networks. Table 3.1 shows a taxonomy of different types of VNFs and their representative examples. These services range from traditional telecom devices to security appliances. In the trend of microservices, the VNFs will become small and modular. New services can be built by composing multiple microservices together in a flexible way.

The objective of NFV is to separate software that defines the VNF from hardware and generic software that creates a generic hosting network functions virtualization infrastructure (NFVI), which executes the VNF. The VNFs and the NFVI are then separated from each other. Compared to the cloud architecture described in the previous chapter, the NFV architecture's mapping can be shown in Figure 3.1. The NFVI maps to both the hardware and infrastructure layer. The VNFs map to the software as a service layer.

In the past 5 years, NFV has grown from research proposal to real deployment. There have been various industry activities to make it prosper. The ETSI consortium has published the NFV architecture documents and has formed various working groups to study the design

Table 3.1 Types of VNFs and examples.

Edge devices	vCPE, IP Edge, BRAS
Gateway functions	IPsec, SSL, NAT
NGN signaling	IMS, VoIP, HSS
Application optimization	CDN, cache, video transcoder, HTTP header enrichment
Performance improvements	Load balancer
Service assurance	SLA monitoring, testing
Mobile network equipments	HLR, SBC, MME, SGSN, GGSN, RNC, eNodeB
Security functions	DPI, Anomaly detection, Firewall, IDS, IPS

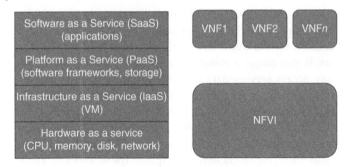

Figure 3.1 NFV architecture in cloud computing stacks.

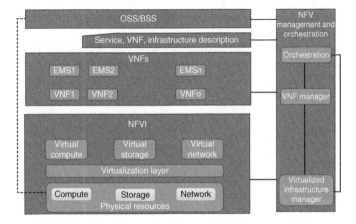

Figure 3.2 ETSI NFV reference architecture.

of different aspects of NFV. Figure 3.2 shows the ETSI NFV reference architecture [50]. It includes the following components:

- *NFVI*: The NFVI provides the technology platform with a common execution platform for a variety of VNFs. It consists of a distributed set of NFVI nodes in various locations to support the locality and latency requirements of the different use cases. NFVI can be categorized into three domains: the compute and storage domain, the hypervisor domain, and the infrastructure network domain. The compute domain refers to the computing hardware in Figure 3.2. The hypervisor domain includes both virtual computing and virtualization layer. The infrastructure network domain refers to the virtual networks as well as the network hardware.

- *NFV orchestrator*: The objective of NFV is to separate the VNFs from the infrastructure, and this includes their management. This management and orchestration block is often complex and composed of great many distributed component parts. On the one hand, it handles the network-wide orchestration and management of NFV resources. It is an integral and essential part of the NFV framework and is specified within the GS NFV management and network orchestration (MANO) documentation. On the other hand, it is used to realize the NFV service topology on NFVI.
- *VNF manager(s)*: It is responsible for VNF life cycle management, including VNF creation, resource allocation, migration, and termination.
- *Virtualized infrastructure manager(s)*: It is responsible for controlling and managing the computing, storage, and network resources, as well as their virtualization.
- *OSS*: Similar to OSS in traditional cellular network infrastructure, the main role of OSS in NFV is supporting the comprehensive management and operations functions.
- *EMS*: It performs the typical FCAPS functionality for a VNF. The current element manager functionality needs to be adapted for the virtualized environment.

NFV is tightly related to various new technologies, including the network virtualization, SDN, and cloud computing. In the next 5 years, NFV will continue to grow and will have a wide deployment in telecom networks. For example, AT&T has announced Domain 2.0 project, which will fundamentally transform their infrastructure using NFV and SDN technologies in the next 5 years. NFV will also have broad applicability in large enterprise networks to reduce their CAPEX and OPEX. We discuss four key technologies of cloud computing that are used in NFVI. The first one is on-demand, self-service, meaning that the operators can easily allocate resources such as server time, storage, and network. The second feature is a flexible and board network access, meaning that any VNF can route traffic to another VNF freely if needed. The third feature is resource pooling. Within a single NFVI point of presence (PoP), one type of VNF may need to deploy on multiple instances. To support dynamically changing resource allocation, VNFs can be deployed on a pool of available resources. The last feature is rapid elasticity for the sake of load surge, failover, and power optimization.

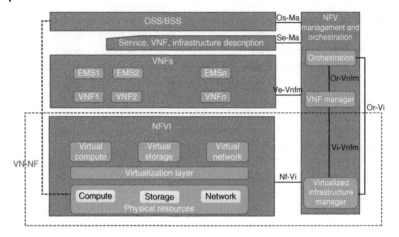

Figure 3.3 OPNFV architecture.

The open-source community OPNFV has been growing and pushing the open-source development of the NFV infrastructure. One open-source NFV activity is called OPNFV [13]. Figure 3.3 shows the OPNFV architecture. It provides the best carrier-grade NFVI, virtualized infrastructure management (VIM), and APIs to other NFV elements. It forms the basic infrastructure required for VNFs and MANO components.

3.2 NFV Use Cases and Examples

ETSI working group has defined a few use cases for NFV in [10]. In total, there are nine use cases. We introduce the high-level description of each use case as follows. More discussions related to telecom and 5G use cases are covered in later sections.

1) *NFV Infrastructure as a Service (IaaS)*: Due to the geographic distribution needs and cost reduction benefits, there is a need for operators to run their network functions on the infrastructure that belong to service providers. For example, instead of deploying its own infrastructure in Europe, a US carrier can choose to deploy its VNFs on a European network service provider's NFVI. In this model, the latter is offering its NFVI as a service.

2) *Virtual network function as a service*: Purchasing and maintaining their own dedicated network service appliances may be too expensive for some small enterprises. Today, the enterprises usually have to buy a multifunction access gateway, which is hard to upgrade or maintain. NFV allows them to purchase functions whenever and wherever needed as a pay-as-you-use model.

3) *Virtual network Platform as a Service (PaaS)*: Service provider today already provides various virtual private network services to their customers. Enterprises need to deploy virtual network to connect their remote offices and employees in a seamless manner. With NFV, the virtual network platform includes not only the routing and switching gear but also various network functions, which can provide additional security and QoS to the enterprise network.

4) *VNF forwarding graphs*: A VNF forwarding graph refers to a pipeline of VNFs that traffic should traverse through. It is also known as SFC. The pipeline can provide a flexible and configurable set of services to a given traffic aggregate. With the VFN forwarding graph, the customer can choose any combination of the services according to the policies from tenants, applications, and networks.

5) *Virtualization of mobile core network and IMS*: Mobile core network needs resource elasticity to accommodate unforeseen demand surge or unpredicted failures. For example, when disaster occurs, the mobile core network faces tremendously increasing calls with reduced resources. Moreover, most mobile network consists of many geographically distributed areas, that is, PoPs or central offices. Virtualizing the core network facilities such as EPC or IP multimedia subsystem (IMS) can increase the capacity of a mobile core network dynamically.

6) *Virtualization of mobile base stations*: In recent years, the virtualization of radio access network has been proposed and drawn significant attention. After virtualizing base stations, multiple service providers can share the same physical resource so that the coverage and resource utilization are both increased.

7) *Virtualization of home environment*: The residential gateway or home gateway is another place where traditional network services are deployed as dedicated hardware. When upgrading the home gateway, often a new hardware needs to be shipped together with

manual installation. Debugging is challenging as the end users do not have the expertise and knowledge to perform troubleshooting. Typical network services in a home environment include parental control, DPI, firewall, and so on. With NFV, the gateway can simply be a forwarding device while other complex functions can be run in the providers' data center. This can help reduce the operational overhead and ease the management tasks.

8) *Virtualization of CDN*: Content delivery network (CDN) hosts massive amount of objects, ranging from web to video, in a distributed manner. The demand of the CDN storage can increase drastically with the content's popularity, such as flash crowd. A popular program's live streaming can put large pressure on both the CDN and the network. Today, CDNs are more and more tightly integrated with the service provider's network given the latter's wide geographic distribution and service-level agreement (SLA) provided. Virtualizing the CDN nodes and deploy them in the cloud can provide resource elasticity as demand changes.

9) *Fixed access network functions virtualization*: The main fixed line access network technologies today are based on digital subscriber line (DSL), for example, VDSL or FTTdp. These technologies require a dedicated electric system to be deployed on the street close to home. These devices are usually expensive as they need to tolerate all sorts of extreme environmental conditions. Access network functions virtualization can greatly simplify these remote devices.

Among all these use cases, 1–4 are in the cloud computing area, 5 and 6 are in the area of mobile computing, 7 is for data center, and the last two are in the area of residential and access networks. Together they demonstrate the wide usage of NFV in today's network.

Virtualizing NFs could potentially offer many benefits including, but not limited to, the following:

- Reduced equipment costs and reduced power consumption through consolidating equipment and exploiting the economies of scale of the IT industry.
- Increased speed of time to market by minimizing the typical network operator cycle of innovation. Economies of scale required covering investments in hardware-based functionality are no longer applicable for software-based development, making feasible other modes of feature evolution. NFV should enable network operators to significantly reduce the maturation cycle.

- Availability of network appliance multiversion and multitenancy, which allows the use of a single platform for different applications, users, and tenants. This allows network operators to share resources across services and across different customer bases.
- Targeted service introduction based on geography or customer sets is possible. Services can be rapidly scaled up/down as required.
- Enables a wide variety of ecosystems and encourages openness. It opens the virtual appliance market to pure software entrants, small players, and academia, encouraging more innovation to bring new services and new revenue streams quickly at much lower risk.

3.3 NFV Challenges

Carrier-grade properties: The telecom service providers have high requirement on the performance, scalability, fault tolerance, and security on the solutions.

- *Efficiency*: The NFV platform should provide the tight NF SLAs on performance or availability, identical to the SLAs offered with dedicated services. For example, the SLA may specify the average delay, bandwidth, and the availability for all the services provided to one customer. To support the SLA compliance, the platform should closely monitor the performance for each customer and dynamically adapt the resources to meet the SLAs.
- *Scalability*: The platform should support a large number of VNFs and scale as the number of subscribers/applications/traffic volume grow. The ability to offer a per-customer selection of NFs could potentially lead to the creation of new offerings and hence new ways for operators to monetize their networks.
- *Reliability*: The platform should abide by NFV reliability requirements. Service availability, as defined by NFV, refers to the end-to-end service availability that includes all the elements in the end-to-end service (VNFs and infrastructure components).

Elasticity: Building on top of the virtualization technology, an NFV platform should be able to leverage the benefit of running instances in the cloud: multiplexing and dynamical scaling. For multiplexing, it allows the same NF instance to serve multiple end users in order to maximize the resource utilization of the NF. On the other hand, for dynamical scaling, when the demand changes, the network operators should be able to dynamically increase/decrease the number and/or size of each NF type to accommodate the changing demands. This, in turn, will allow the telecom service providers to offer their customers with the "pay as you grow" business models and avoid provisioning for peak traffic. It should support subscriber-based, application-based, device-based, and operator-specific policies simultaneously. Moreover, adding or removing new NFs should be easily manageable by the network operator, without requiring the physical presence of technicians on the site or having the enterprise customers involved. It should also be possible to accurately monitor and reroute network traffic as defined by policy.

Openness: Aligned with the Open NFV strategy in the HP NFV business unit, the NFV framework should be capable of accommodating a wide range of NFs in a nonintrusive manner. It should support open-source-based and standard solutions as much as possible, meaning that the NFs should be implemented, deployed, and managed by operators, enterprises, or third-party software vendors.

3.4 NFV Orchestration

NFV management and orchestration component, that is, MANO, is an integral piece of NFV architecture and a critical component to the success of NFV. It is responsible in managing all the resources for NFV framework; not just the networking resource but also the compute and storage resources. According to ETSI's MANO working group, it contains three key components: the NFV orchestrator, VNF manager, and virtualized infrastructure manager (VIM).

- NFV orchestrator is responsible for the installation and preparation for new services, which is also called the network service onboarding process. It handles the entire life cycle of the new network service, including resource allocation, validation, and authorization. While the VNF manager handles the life cycle of each VNF instance, the NFV orchestrator operates at a higher level. Its functionality includes registering a network service in the catalog, and onboarding the network service, instantiating the network service, scaling up/down, managing VNF forwarding graph related to the network service, and finally terminating the service.
- VNF manager, as the name indicates, handles the life cycle of each VNF instance. It includes the configuration, preparation,

and running of the VNF instance, monitoring its healthiness and interacting with other MANO components on behalf of the VNFs. The actual tasks of VNF manager include the following: instantiate a VNF, monitor its resource utilization, scale up/down the VNF, update or upgrade, and finally terminate the VNF and release all its resources.

- VIM controls and manages the NFVI components, to better accommodate the need for various network services. For compute resource, it manages the physical resources and the virtual resources, for example, servers, VMs, CPUs, and memories. For networking, it controls the network devices, links, routing, addressing, and QoS.

There are various open-source activities on MANO development. SDN controller is considered an integral part of MANO to control the network resources. Besides the three main components, there are other aspects such as fault tolerance, policy management, security and privacy. In the following, we discuss two interesting problems in the NFV orchestration area.

3.4.1 NFV Performance Characterization

Before deploying VNF to the production network, the service provider needs to extensively test and validate the VNFs given the high requirement on availability and performance. However, today there is no way for the service provider to prevent or detect the performance degradation caused by third-party VNFs. Similar to other applications, the underlying hardware server characteristics have a deep impact on the performance. Parameters such as processor architecture, clock rate, memory channels and speed, memory latency, and bandwidth of interprocessor buses have a strong impact on the performance of the specific application or VNF running on that HW. In NFV, we need to allocate resources to not only one single server but multiple servers. Scaling to a huge number of devices, connections, and services is indeed a challenge. Every component of the system will need to scale up and scale down with demand variations. This is necessary for efficient operation and obtaining the cost savings required. But, how much resources to give to each VNF? Answering this question is not easy. Different VNFs may have different requirements on compute and network resources. For example, while firewalls are bounded by network throughput others such as load balancer may be bounded by network and compute for session state management. Providing additional CPU to an I/O bound VNF is not helpful.

Characterizing various VNFs and understanding how they scale with different kinds of resources is a key step toward answering the above questions. Research has been proposed to develop tools to determine the virtualization setups and configuration options to optimize VNF performance and automatically scale the VNF resource allocation with workload [51]. One difficulty is that there is a large number of configuration knobs and hardware settings, for example, CPU pinning, c-states, and memory interleaving; they are all available in NFV deployments. We develop efficient search algorithms for this. This type of performance characterization tool can significantly improve the onboarding process, giving more confidence to the network operators to support this area. It can help NFV infrastructure providers to better understand, control, and manage the VNFs.

Benchmarking the VNF's network and compute performance in real NFV deployment is important for the resource allocation. Figure 3.4 shows a typical VNF performance characterization process. The testing is performed for each configuration. Different virtualization technologies, such as Intel DPDK, SR-IOV, and VT-d, are configured depending on the scenarios. Tests are done with different VM placement and configurations. VMs may be deployed on the same hosts or across hosts, both have different impact on the real performance. Within each server, different core scheduling

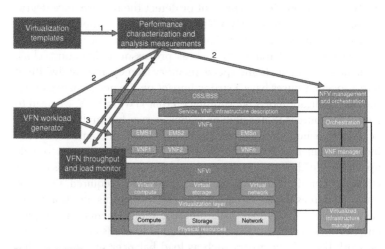

Figure 3.4 VNF performance characterization.

algorithms can be tried, such as core pinning and adaptive core scheduling. Different deployment scenarios can also be a configurable option, for example, deploying on a single powerful VM or on multiple VMs. Finally, to reveal the actual performance that one will experience in the real network, we need to test with different network traffic, not only using plain dummy traffic to test throughput but also application-aware traffic.

3.4.2 NFV Performance Improvements

Telcos have transitioned most of their communications to standard IP networks and are now starting to migrate most of their computing to industry standard servers. In addition, with the advent of 5G technology, there is a push toward developing novel real-time services in the areas of augmented reality, autonomous driving, and IoT. In NFV environments, monolithic complex network functions running on specialized hardware are decomposed into smaller functional units and dynamically orchestrated onto a virtualized cloud and edge infrastructure. These network functions include both control plane processing (signaling, authentication, etc.) and data plane processing (routing, firewalling, etc.) and are encapsulated in a VM as a VNF. Service chaining defines routing instructions that direct network packets through a pipeline of VNFs. A core component of the NFV infrastructure is a highly agile network that enables efficient and dynamic service chaining.

While the NFV transformation is under way, another trend is the move toward using large pools of compact, power-efficient compute nodes/SoCs that are integrated with large stores of fast persistent memory and custom high-speed interconnects. These computing architectures span from rack scale to datacenter scale and are expected to impose unprecedented scale, bandwidth, and latency requirements on future data center networks. This architecture is usually composed of a large number of nodes; each node is composed of a tightly coupled set of processor cores and some local memory. Each node acts like a traditional computer and it has its own operating system. All the nodes are connected using a high-speed memory fabric, called next-generation memory interconnect (NGMI). Such a memory fabric with high bandwidth and low latency could be used to speed up internal communications of NFV applications and expedite the packet passing between VNFs. It has a large pool of CPUs and memory and enables a flexible binding of memory to CPU. This can provide better load adaptation, where computing can be dynamically reconfigured to scale up or down based on the network traffic load and patterns seen in telco networks.

3.5 NF Modeling

Given the widespread deployment of proprietary network functions (NFs), many network management functions (e.g., testing, policy compliance, and vendor interoperability) need accurate models of NFs to ensure correctness and reliability. However, due to their proprietary and complex nature, there is a lack of standardized behavior model of these NFs. In the following, we review existing NF models and survey several approaches to generate NF models.

In the NFV architecture, each NF is often viewed as a black box that processes traffic arbitrarily. However, as more NFs are developed by diverse vendors and deployed in richer scenarios, the "box" view of NFs has already hindered its more advanced application and development. For example, if NFs are viewed as boxes without knowing its internal logic, there is no way to reason about their forwarding behaviors; thus, NFV operators cannot troubleshoot if they are misbehaving and cannot verify network-wide policy compliance such as reachability. Knowing NF's behavior can also help monitor the policy compliance by generating meaningful testing traces. Thus, having a forwarding model of individual NF that describes its internal forwarding logic is important for NFV network management tasks.

Figure 3.5 shows the important role that NF models play in the whole NFV ecosystem. NF models connect NFV applications and NF implementation. On the one hand, from NFV operators' view, several applications on top of NFV platforms are based on NF models. For example, when an operator needs to generate test traces for deployed NFs, the operator needs to assume NFs' behaviors (i.e., model) to completely cover the test space; when the operator verifies network-wide properties (e.g., reachability), whether a flow gets through an NF needs to be inferred from the NF model.

Existing works usually build NF models according to their application scenarios, so that the models are tailored specific to their applications and not generic. They fall into two categories. The first category aims to build models used for network management tasks (e.g., testing, traffic engineering, and migration), and the second one focuses on improving NF software, including its performance,

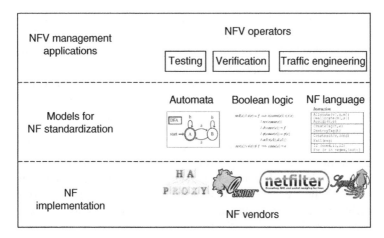

Figure 3.5 NF model overview.

reliability, and robustness. The representative modeling approaches are as follows:

- *High-level box view*: In most existing NFV platforms and NF performance optimization solutions, an NF is viewed as a black box with traffic traversing it, but the NF's internal logic is not differentiated. In practice, an NF is implemented as a VM with the NF software running inside it, the VM is deployed on general servers, allowing traffic to be routed through. This coarse-grained view only supports simple operations: NF bootup, migration, suspension, and destruction.

  OpenNF [52] takes one step further; it recognizes an NF as some running logic plus a state blob. The state blob can be created/destroyed/moved between NF instances and encoded/decoded by NFs. This abstraction allows more flexible flow-level operations (i.e., flow migration between NF instances), but is still limited to flow management only. Following the requirements to identify the state blob, StateAlyzr [53] proposes an approach to identify state variable in NF programs. It basically uses some NF-specific features (e.g., program structure, packet libraries), and leverages program slicing techniques to find out state variables. This work provides an approach to analyze NFs but does not go further in NF modeling.

- *Manually generated models*: A set of works view NFs as deterministic finite automata (DFA). For example, BUZZ [54] uses DFA to represent each NF in the network, finds possible violation states in each DFA, and use a network-wide model to generate traces (that may trigger the violation) for testing. In another example, SFC-Checker [55] also views NFs as DFAs; it mainly answers a question, given a network with NFs and a series of packet traces, whether there is a risk of violations (unexpected reachability/unreachability). The DFAs in these works are manually generated and directly used to replace the NFs, which implies possible incompleteness.

- *Modeling language design*: SymNet [56] proposes a symbolic execution-friendly language to remove these two features for NF programming. Both works aim to develop NF programs that can be symbolic executed so that program bugs can be found out early during the development. These two approaches provide new ways to model NFs but still require programmers to writing NF programs.

3.5.1 Source-Code-Based Modeling

One way to perform modeling is to leverage recent advances in code analysis techniques. Wu *et al.* [57] analyzes the source code of a given network function to automatically generate an abstract packet forwarding model. It is built on top of the following code analysis methods.

3.5.1.1 Background

Program slicing

A program slice is a minimum set of statements in a given program that lead to certain behaviors. The behavior is expressed as the values of variables in a certain statement. Program slicing has been shown to be useful for program debugging, parallelization, and integration [58].

The basic methodology to get a program slice is to analyze the dependency between program statements. Within one statement in a given program, the value of the left-hand-side (LHS) variable depends on that of the right-hand-side (RHS) variables; and between statements, the value of an RHS variable in a statement depends on the preceding statements where that variable is on the LHS. This dependency analysis results in a "static" program slice where all statements in that slice *might* lead to the final behavior. A "dynamic" program slice is all statements that *really* lead to the final behavior, which requires execution analysis based on actual variable values [59]. Several research projects have improved program slicing techniques – for example, interprocedure slicing and system dependency analysis [60, 61].

Symbolic execution

Symbolic execution is another program analysis approach that substitutes one or several program variables by symbolic values. As the program runs, whenever a branch instruction is met, the program is forked and both branches proceed. The program state variables and path constraints are kept by each path individually. At the end of a path's execution, concrete values of symbolic variables are computed according to the path constraints [62, 63].

Symbolic execution can exercise all possible execution paths, but path explosion can happen with a large code base. Various efforts have made symbolic execution practical in NF verification. To reduce the branching factor (number of branches at each branch instruction), Dobrescu and Argyraki [64] and SymNet [56] propose to write NF programs in a style with bounded loops and data structures, and BUZZ [54] constrains the number and scope of symbolic variables.

NF state analysis

OpenNF [52] is a flow state management framework that enables joint control with SDN and NFV frameworks. An important issue when integrating existing NFs into OpenNF is to identify state variables in NF programs. StateAlyzer [53] defines features of state variables in an NF program, leveraging several program analysis techniques (e.g., program slicing, system dependency analysis) to identify them. These features are listed as follows:

- *Persistent*: The variable has a lifetime longer than the packet processing loop.
- *Top level*: The variable is actually used during packet processing.
- *Updateable*: The variable's value is updated during packet processing, that is, usually an LHS variable.
- *Output-impacting*: The variable impact variables in the packet output function.

These features and categorization of variables are used for code analysis to generate the NF models.

3.5.1.2 Modeling Example

We use a Layer-4 load balancer as an example of the NF source code and illustrate how the above techniques can be used to synthesize

```
1  from scapy.all import *
2  # Constants
3  ROUND_ROBIN = 1
4  MTU = 1500
5  # Configurations
6  mode = ROUND_ROBIN
7  LB_IFACE = "eth0"
8  LB_IP, LB_PORT = "3.3.3.3", 80
9  servers=[("1.1.1.1", 80), ("2.2.2.2", 80)]
10 # Output-Impacting States
11 f2b_nat, b2f_nat={}, {}
12 rr_idx = 0
13 cur_port = 10000
14 # Log States
15 pass_stat, drop_stat=0, 0
16 # callback function
17 def pkt_callback(pkt):
18    global drop_stat, pass_stat, rr_idx, cur_port
19    si, di = pkt[IP].src, pkt[IP].dst
20    sp, dp = pkt[TCP].sport, pkt[TCP].dport
21    if dp == LB_PORT: # pkt from client to server
22       cs_ftpl, sc_ftpl = (si,sp, di,dp), (di,dp, si,sp)
23       if cs_ftpl not in f2b_nat: # new connection
24          if mode == ROUND_ROBIN:
25             server = servers[rr_idx]
26             rr_idx = (rr_idx+1) % len(servers)
27          else: # Hash to a backend server
28             server = servers[hash(si) % len(servers)]
29          n_port = cur_port
30          cur_port+=1
31          cs_btpl = (LB_IP, n_port, server[0], server[1])
32          sc_btpl = (server[0], server[1], LB_IP, n_port)
33          f2b_nat[cs_ftpl], b2f_nat[sc_btpl]=cs_btpl, sc_ftpl
34          nat_tpl = cs_btpl
35       else: # existing connection
36          nat_tpl = f2b_nat[cs_ftpl]
37    else: # pkt from server to client
38       sc_btpl = (si, sp, di, dp)
39       if sc_btpl in b2f_nat:
40          nat_tpl = b2f_nat[sc_btpl]
41       else: # no initial outbound traffic is allowed
42          drop_stat+=1
43          return
44    pass_stat+=1
45    pkt[IP].src, pkt[TCP].sport = nat_tpl[0], nat_tpl[1]
46    pkt[IP].dst, pkt[TCP].dport = nat_tpl[2], nat_tpl[3]
47    for f in fragment(pkt[IP], fragsize=MTU-len(Ether())):
48       sendp(Ether()/f, iface=LB_IFACE)
49 def LoadBalancer():
50    sniff(iface=LB_IFACE, prn=pkt_callback, filter="tcp")
51 if __name__=="__main__":
52    LoadBalancer()
```

*varaiable format (cs—sc)-(f—b)(s—d)(i—p—tpl): sc—cs is direction between client and server, f—b is a side of frontend/backend, s—d is source/destination and i—p—tpl is IP, port or 4-tuple.

Figure 3.6 Load balancer code and a slice (bold and bold italics).

NF models. Figure 3.6 is its implementation based on the model in [65]. The high-level logic of the code is as follows. Inbound packets that have the IP addresses/ports of clients and the load balancer would be mapped to the IP address/ports of the load balancer and servers. If an inbound packet is a new flow never seen before, one of the backend servers is picked for the mapping and the mapping is stored; otherwise, the mapping is read from the dictionary and used for the address/port translation. For the outbound packets, the packets of existing flows would have address/port translation in the same way, but outbound packets of a nonexisting flow would be dropped (i.e., only inbound packets can initiate address/port translation mapping).

From reading this code, we gain important insights about NFs. First, different configurations can lead to different program behaviors. For example, the variable "mode" is used to configure how a backend server is selected for a new flow, and it can be either round-robin or random hash. Some existing NF models [65] fail to capture this detail. Second, state variables are dynamically updated as packets are processed. It causes the action on packets to be different at runtime. In this example, whether a flow's 4-tuple is stored in the dictionary is a state that causes the actions on the flow's first packet and that on the remaining packets to be different. Third, NF programs naturally have two key pieces of logic, that is, packet processing and state management. Thus, we could use program analysis to refactor the code and identify the minimum set of statements that capture such forwarding logic. For example, in Figure 3.6, the bold and bold italics lines are a (dynamic) program slice where the load balancer relays the first packet of a flow. This small code snippet captures the forwarding behavior of this NF. Identifying this snippet automatically from the original code would improve the efficiency of manual analysis and automatic verification.

Refactoring the logic in an NF program could benefit NF verification in many ways. For example, finding out the state update logic helps to build a finite state machine (FSM) of that box; in the network-wide, assembling FSMs of multiple NFs can help to find out the network-wide invariant violation. In addition, refactoring NF logic helps to minimize the number of statements that would lead to a certain behavior (e.g., packet corruption, unexpected drop), and this minimization not only eases manually program analysis but also speeds up other automated program analysis solutions (e.g., symbolic execution).

3.5.1.3 Models

Referring to various existing NF models [54–56, 65–67] and the current hardware programmability [68–70], we adopt an OpenFlow-like model with a stateful data plane extension. The abstract model is expressed as tables in Figure 3.7, and each table describes the packet processing logic under a certain configuration (e.g., c_1 in the figure).

Match		Action	
Flow	State	Flow	State
Configuration = c_1			
f_1	$s_1 \wedge P(f_1, s_1)$	$Fwd(f_1, s_1)$	$Upd(f_1, s_1)$
...	...	...	...
Configuration = c_2			
f_2	$s_2 \wedge P(f_2, s_2)$	$Fwd(f_2, s_2)$	$Upd(f_2, s_2)$

Figure 3.7 Model.

Each entry in the table represents certain processing logic and consists of match and action fields. As a stateful data plane, the match/action fields operate on both flows and state variables. The match is executed on flows and states, and the action not only forwards packets (with possible transformation) but also triggers state transition.

In table c_1 in Figure 3.7, if an incoming packet matches a flow pattern f_1, the internal state is in s_1 and a predicate $P(f_1, s_1)$ is satisfied, then the packet is sent out with possible transformation $Fwd(f_1, s_1)$ and the internal state is transited to $Upd(f_1, s_1)$. In our running example code, each frontend incoming packet would be checked to determine whether the flow was seen before (in the address/port translation mapping), which is the predicate of flow and states in the match field; the first packet of each new flow triggers address and port translation (stored in a dictionary), which is a state transition; and sendp() is an action on packets.

3.5.1.4 Model Extraction Overview

In the following, we sketch a white box modeling approach based on our existing work. In this approach, we assume that the source code of the NF program is known and propose a modeling framework called Lancet. An NF program is assumed to have a processing loop and packet input/output functions, and an NF model is predefined like a stateful match action table. This work is published as NFactor [57]. The process is shown in Figure 3.8. It takes three steps as follows:

1) It first conducts backward slicing from packet output function to get packet slice and from state variable assignment statements to get state slice.
2) It then symbolically executes the union of both slices and get multiple execution paths. For scalability, we need to constrain the value of some variables.

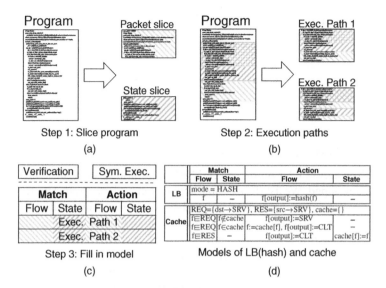

Figure 3.8 Source-code-based NF modeling method. (a) Step 1: slice program; (b) Step 2: execution paths; (c) Step 3: fill in model; (d) Step 4: Models of LB(hash) and cache.

3) Finally, each path is filled into the stateful match action table as a row, conditional statements about packets/states are put into flow/state fields of match column, and intersection of an execution path with packet/state slice is put into flow/state fields of action column.

This approach gives an accurate model but is constrained by the prerequisites that source code is known. Figure 3.8(d) shows the modeling results of running NFactor on a load balancer HAProxy [71] and a cache server WWWOFFLE [72]. The logic in the match action table is described in SNAP language for abbreviation. HAProxy is a load balancer and configured in hash mode, and it randomly chooses one of its backend servers to forward the incoming requests. WWWOFFLE is a cache. When it observes a request for the first time, it forwards the request to the server; when a response is back, it records the content of the response and forwards it to its client side; and when the cache observes a request whose response appears before, it immediately replies the client with the response without forwarding the request to the server.

Models, if viable, can faithfully rely on code analysis.

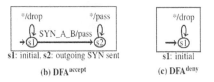

(b) DFA^{accept} (c) DFA^{deny}

3.5.2 Modeling Applications

With an accurate NF model, several NFV applications can be designed and developed. We list the example applications: NF program generation, stateful network verification, and network policy composition.

Verifying correctness of a network's configurations ahead of its actual deployment can avoid runtime errors. There a several stateless or stateful network verification solutions [55, 67, 74, 75] aiming to verify network properties such as reachability and isolation. Our model provides the NF model to support these solutions.

We use SFC-Checker [55] as an example. SFC-Checker models each NF as a DFA and composes NF DFAs in a network to check reachability. Thus, we transform the model into DFA as the foundation to apply SFC-Checker.

In this section, we use examples of IDS and load balancer to illustrate the modeling process. It induces NF programs [76, 77] to the abstract models in Figure 3.8.

- An IDS is configured with a list of denied flows "DENY," and flows not in this list are allowed by default. An IDS has no output-impacting states.
- A load balancer in round-robin mode forwards new flows to one of the backend servers pointed by an internal state "index." And the index increased circularly for each new flow.

To generate a DFA, one needs to decide the state space and the state transition. Each entry in the model can express a state transition as follows.

The packet processing logic takes f as input and outputs forward(f,s) and the states transit from s to transit(f,s).

However, the state space needs to be in proper granularity. Too fine a granularity would cause a state space explosion. For example, a cache can have 2^{cache_size} possible values, which is not practical to verify. While too coarse a granularity cannot guarantee the correctness or efficiency of verification. For example, the state "idx=*" in load balancer can be used to represent any states but is not useful to identify the exact next hops.

Therefore, we propose to *refine each entry to the granularity of possible next hops*. For example, if the load balancer has two backend servers configured, then the state-matching field is refined to two states "idx=0" and "idx=1" to indicate two possible next hops.

Thus, we can transform the model to DFAs (as is shown in Figure 3.8). Next, we can leverage SFC-Checker [55] to check the end-to-end reachability, and the result is shown in the evaluation section.

Model-based NF programming is the reverse process of model generation. To generate an NF program from a match action table, the program first contains a loop starting with packet input function. Each row in the match action table is transformed to a branch in the program. The branch has the conjunction (AND) of the state and flow fields in the match column as its condition statements and performs the state/flow statements in action column. This program can be viewed as a microservice with the same functionality with the NF, focusing on the logic of NFs. The challenge in model-based NF programming is to guarantee other program requirements, including performance, isolation, and providing management API.

The DFA model of NFs lays the foundation of several stateful network verification solutions. While we list two existing DFA-based solutions in previous sections, there are still challenges in stateful network verifications. For example, how to guarantee the completeness and soundness of the verification? How to efficiently verify large networks with frequent changes? We are abstracting and solving these problems based on DFA theory.

NF models can also be used in service policy composition. When deploying a network-wide policy composed by several NFs, there may be conflicts between NFs. For example, deploying a cache and a firewall in different orders would lead to different results, for example, one client's content may be cached and provided to another client, which was intended to be blocked by the firewall. NF models can be used to determine this order in service policy composition; the operator models the input/output of each NF when chaining them, so that the end-to-end requirements can be guaranteed.

3.6 VNF Placement

VNF placement problem entails choosing where to place instances of VNFs on servers in a physical network in order to accommodate the traffic for a set of service chains. A few examples of VNFs are firewall services, intrusion detection, caches, and proxies. Each service chain is a stream of network packets flowing through a sequence of VNFs at a certain rate. Network traffic for a given service chain must visit the chain's sequence of VNFs in the specified order. In the VNF placement problem, one must place (possibly multiple) instances of each VNF on servers and choose the route(s) for each service chain in such a way that the physical network can accommodate the traffic for all service chains or, if not all can be accommodated, the highest priority service chains. The traffic for a given service chain may be split among multiple paths in the network when multiple instances of a specific VNF are used. Moreover, the network onto which we are placing VNFs may have heterogeneous server types, such that one server type may be more efficient at running a given VNF than others. Server heterogeneity must be taken into account when choosing where to place VNF instances. VNF placement is a challenging combinatorial problem because it involves a large number of discrete placement decisions. This class of decision problem is known to be NP Hard.

The VNF placement problem resembles that of virtual network embedding (VNE), in which one must determine how to place a set of virtual networks onto a physical substrate network. In one respect, each service chain in a VNF placement problem can be viewed as a virtual network to be mapped onto a physical network. However, VNF placement differs in that each node (VNF) in a virtual network can be mapped to multiple instances that are placed on different nodes in the physical network. VNE approaches do not address how to map each VNF into instances. VNF placement combines the problems of mapping service chains into virtual networks and then maps the resulting virtual networks into the physical network.

There are multiple objectives when placing VNFs. A service provider may want to host VNFs on as few servers as possible in order to minimize operating costs and leave open servers for future VNF needs.

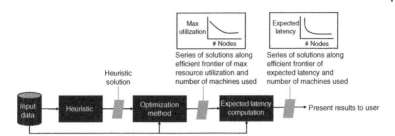

Figure 3.9 VNF placement system overview.

At the same time, he may want to ensure low network latency for his customers. These two objectives are in direct conflict, in that the former involves concentrating traffic in the network, whereas the latter implies spreading traffic out to avoid network congestion.

In the following, we introduce a work that uses a mixed integer programming (MIP) model that explicitly captures the effect of network traffic on latency while maintaining a linear model; we minimize the maximum utilization over resources in the network. Minimizing the worst-case utilization avoids the situation in which a small number of congested resources induce outsized delays on network traffic.

While MIP can compute optimal solution, it cannot scale to a large size of network. We overcome this challenge by having a two-step process. As illustrated in Figure 3.9, we first propose a fast, scalable round-robin heuristic for VNF placement, which generates an initial VNF placement solution. This initial result is then fed to the MIP to help speed up the search process. Secondly, our MIP algorithm balances the competing objectives of minimizing congestion-induced latency and minimizing the number of servers used. It minimizes a weighted combination of two metrics: (1) the number of servers used to host VNF instances and (2) the maximum utilization over network resources, which we use as an approximation for latency. The optimization method generates multiple VNF placement solutions for different relative weightings of the two objectives, thereby generating solutions along the efficient frontier of number of servers and maximum resource utilization.

We present the model and the optimization formulation as follows:

- *Model parameters*:
  - N: The set of all nodes in the network (servers and switches)
  - $L \subset N$: The set of servers, which are leaves in the tree network

- V: The set of VNF types. Instances of these VNF types must to be assigned to servers in the physical network in order to accommodate service chains
- C: The set of service chains to be mapped to the network
- $\beta \in [0, 1]$: A parameter representing the relative weight between two metrics, number of servers used and maximum utilization, in the objective function.

- *Decision variables:* The decision variables describe the assignment of VNF instances to leaf nodes, the mapping of each service chain to one or more paths in the network, the volume of flow for each chain along each of its paths, the rate of traffic into each node, and performance metrics associated with the solution.
  - $x_{v,l} \in \{0, 1\}$ indicates whether an instance of VNF type v is placed on leaf l.
  - $y_{i,l}^c \in [0, 1]$ is the fraction of traffic for the ith function in service chain c that is served by leaf node l.
  - $z_{i,k,l}^c \in [0, 1]$ is the fraction of traffic going from the ith to $(i + 1)$st function in service chain c that travels from leaf node k to leaf node l.
  - $b_k \geq 0$ is the total traffic rate in packets per second into node $k \in N$.
  - ρ is the maximum node utilization over all nodes in the network.
- *Constraints:* The MIP's constraints ensure that flow for each service chain is conserved at each node, that the solution does not use more than the available network resources, and that the maximum utilization metric is measured.
- *Model objectives:* The objective is to minimize a weighted combination of the number of nodes utilized and the maximum utilization over all nodes in the network.

$$w = (1 - \beta)\rho + \beta \frac{1}{|L|} \sum_{v \in V, l \in L} x_{v,l} \tag{3.1}$$

When $\beta = 0$, the objective reduces to minimizing the maximum utilization over all nodes in the network. This choice of objective has the effect of distributing the traffic as uniformly as possible in order to reduce the highest utilization over all nodes. If instead $\beta = 1$, the objective becomes minimizing the total number of nodes used to host VNFs. A placement that minimizes the number of VNFs tends to concentrate traffic in part of the network, leaving other network resources unused. Solving the MIP over a range of $\beta \in [0, 1]$

yields a set of solutions that represent different trade-offs between performance and server usage.

- *Solution procedure:* We solve the MIP multiple times for different values of β in the range $[0, 1]$ to produce a set of solutions along the efficient frontier of maximum node utilization versus number of servers. As β increases, the solutions favor decreasing the maximum utilization ρ over minimizing the number of nodes used. The heuristic solution is used as a starting solution for the first MIP, which helps speed its execution. Then, for each new value of β, the preceding solution can be used as a starting point.

3.7 Summary

In this chapter, we introduce the NFV overview. We first present the NFV architecture proposed by ETSI and its relationship with cloud computing. We then present its use cases in telecom network. We discuss the challenges in virtualizing the NFs and using them in telecom network. Finally, we further explain the difficulties and problems in deploying NFV in reality, in terms of measurement, characterization, and modeling.

4

Software-Defined Networks Principles and Applications

Software-defined networking (SDN) is growing as a solution in dynamic environments where customers need to adapt to new protocols and standards. Different protocols such as OpenFlow have arisen in order to facilitate the implementation and deployment of SDN-enabled networks. Customers are looking for a programmable interface where they can program Ethernet chips to meet their custom needs; with hard-coded silicon and protocol-dependent interfaces, the adaption of new protocols requires several years to develop, implement, test, and deploy. However, customers are also looking to insert SDN vertically in the data path, allowing the SDN application to coexist and function together in addition to traditional forwarding, also facilitate adoption, or solve specific networking challenges.

SDN enables centralized, programmatic, and fine-grained control of network flows, based on a global view of the network state. A logically centralized SDN controller runs SDN applications targeting important network functions such as load balancing/traffic engineering, quality of service (QoS), and fast failure recovery. The SDN controller programs flow rules in the forwarding tables of SDN-enabled network devices, typically using the OpenFlow protocol. As a basic technology, SDN can enable dynamic control and routing in many network environments. For example, modern enterprise networks use SDN applications to provide QoS to latency-sensitive or bandwidth-hungry applications. It can enhance visualization and network troubleshooting. Cloud computing environments use SDN to provide network virtualization capabilities. And finally, NFV infrastructures use SDN to provide dynamic traffic steering, virtual network function (VNF) load balancing, and service function chaining. In this section, we

Network Function Virtualization: Concepts and Applicability in 5G Networks, First Edition. Ying Zhang.
© 2018 John Wiley & Sons, Inc. Published 2018 by John Wiley & Sons, Inc.

focus on the basic concepts and development of SDN itself, and in the following section, we will focus on the service function chaining, that is, the use case of both SDN and NFV.

4.1 SDN Overview

The significant increases in users, devices, applications, and traffic have imposed new challenges to the service providers to reduce the total cost of ownership (TCO) and improve average revenue per user (ARPU) and customer retention. The SDN paradigm arose to solve some of these new challenges. SDN decouples the control and forwarding planes. so that each plane can independently scale in order to reduce TCO. SDN provides a set of open APIs so that network programmability has been introduced. New services and innovation can be enabled. With the flexibility in network control, SDN has been used to provide network virtualization for resource optimization. The OpenFlow concept enables intelligent flow management. In this section, we provide an overview of the SDN technology and business drivers and describe the high-level SDN architecture and principles. We discuss a few scenarios of its use cases in mobile access aggregation networks and the cloud networks. We further introduce more details in SDN controller and data plane protocols.

4.1.1 Motivations

From the business perspective, the main motivation of SDN in service provider networks is those challenges operators face. The three-dimensional business drivers are TCO reduction, increased ARPU, and improved customer retention. These three dimensions are connected and they are important factors determining the way networks are architected and designed. While mapping these business drivers to technology drivers, scaling and virtualization are primarily required to address reducing TCO, while service velocity and innovation is required to address increasing ARPU and customer retention.

From the technical perspective, the motivation can be elaborated in three aspects.

First, it can help scale the control and data plane independently. Given the business driver of SDN is for the network to catch up with the demand. Independent scaling can help with accommodating the bandwidth growth. Bandwidth growth comes from different types of applications. For example, the dominant traffic video applications have a need for scaling data plane. But the machine-to-machine applications or VoIP traffic calls for scaling control plane. These different types of traffic in next-generation networks would pose independent scaling of control and data plane to meet next-generation traffic growth and demands.

Second, it can help improve the service velocity and innovation. It is imperative that network can be programmable for service providers to use network as a platform to expand their service models to include third-party applications. This helps expand the current business models. Rapid development and deployment of not only in-house services but also third-party applications from other content providers is important for the variety and quality of new network services.

Third, it can enable flexible and efficient network virtualization. When operating different network services, for example, virtual private networks (VPNs), video network, service providers usually deploy an overlay network for each type of network. For example, they have one network for wireline application, another one for wireless, and another for business applications. As we move toward converged networks, the goal is to design one network for many services by slicing this physical network into multiple virtual networks, one for each type of application.

4.1.2 Architecture

Given the significant interest because of the key drivers, an increasing amount of attention has been raised on SDN. However, until now, there has been no consensus on the principles and concepts of SDN. In the following, we take the initiative to summarize the key aspects of SDN from service providers' perspectives and the four main concepts that describe the SDN in service provider networks. Figure 4.1 illustrates the overview architecture and four concepts. Figure 4.2 shows the components in each layer in an SDN architecture.

4.1.2.1 Separation of Control and Data Plane

The separation and centralization of the control plane software from the packet-forwarding data plane is one of the core principles of SDN. This differs from the traditional distributed system architecture in

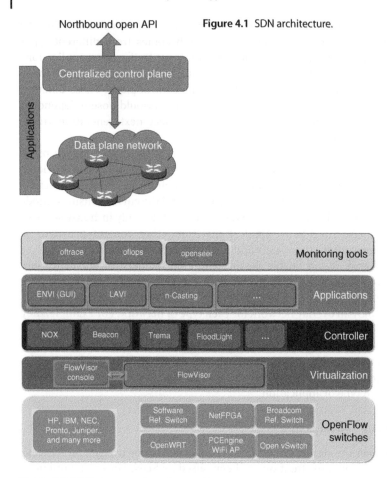

Northbound open API

Figure 4.1 SDN architecture.

Figure 4.2 SDN components and examples.

which the control plane software is distributed across all the data plane devices in the network. Figure 4.3 shows the transition from today's integrated box view to the separation view. In this design, the centralized control enables deployment of new services and applications that are not possible or are very difficult in traditional distributed networks. The deployment of new services is much faster compared to upgrading a whole networking device as it is done by current vendors. The separation of control and data planes enables independent and parallel optimizations of the two planes. We envision that this path

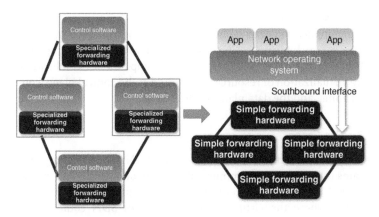

Figure 4.3 SDN comparison.

will lead to highly specialized and cost-effective high-performance packet-forwarding data plane devices and control plane servers, thus reducing the CAPEX of service providers significantly. And finally, since network information, applications, and services are concentrated at a centralized location in SDN, operations such as network orchestration and monitoring become much easier and cost-effective by avoiding individual software upgrades at multiple device locations prevalent in current distributed networks.

4.1.2.2 Northbound Open APIs

Open APIs enable developers to exploit the centrally available network information and the mechanisms to securely program the underlying network resources [78]. They not only enable rapid development and deployment of both in-house and third-party applications and services but also provide a standard mechanism to interface with the network.

4.1.2.3 Southbound Control/Data Plane Protocol

A secure and extensible control plane signaling protocol is an important component for the success of SDN. It should enable efficient and flexible control of network resources by the centralized control plane. OpenFlow [13] is a well-known control plane signaling protocol that is standardized and increasingly being made extensible and flexible. There are other southbound protocols such as OVSDB, which is more backward compatible with existing protocols.

4.1.2.4 Applications

Multiple applications can be built across the control and data plane, such as network virtualization, QoS, service function chaining, and resilience. For example, SDN can be used to implement virtualization, where virtualization of resources is required at the control plane, control channel, and the data plane levels.

4.1.3 Use Cases

We discuss two important use cases of SDN: access/aggregation domains of public telecommunication networks and data center networks. Access/aggregation network domains include mobile-backhaul, where the virtualization principle of SDN could play a critical role in isolating different service groups and establishing a shared infrastructure that could be used by multiple operators. On the other hand, using SDN in data center can improve load balancing and dynamic resource allocation (hence, increased network and server utilization) and the support of session/server migration through dynamic coordination and switch configuration. The decoupling principle of SDN could provide the possibility to develop data-center solutions based on commodity switches instead of typically high-end expensive commercial equipments. We will discuss more about the mobile network use case in a later chapter, so we focus on its data center use cases here.

Data centers are an important area of interest for service providers. Applications such as "cloud bursting" are bringing together private and public data centers like never before. Distributed data centers have been widely used in today's large-scale enterprise networks, such as Google B4 network [79]. In such scenarios, the inter-data center wide area network (WAN) providers become a very important component in efficient and cost-effective connectivity and operation of distributed data centers. Figure 4.4 illustrates this. The static provisioning mechanisms used currently by operators to provision resources on the WAN will not only degrade the performance of distributed data centers but also increase the cost of operation. Thus, it is necessary that the inter-data center WAN evolves into a networking resource that can be allocated and provisioned with the same agility and flexibility that is possible within data centers. SDN is a very good candidate to enable such agile and flexible WAN operations the data center operators are expecting. Centralized control of the WAN

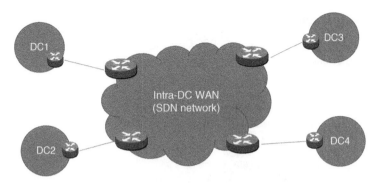

Figure 4.4 Inter-data center WAN: a use case of SDN.

using an SDN controller provides its northbound Open API to enable intelligent traffic engineering and complex policy management. It not only provides a single unified control of the WAN resources but also enables on-demand provisioning of network resources for multiple clients based on their policies and SLAs, in a seamless fashion.

Another advantage of a centralized SDN controller in the inter-data center WAN is that it facilitates unified control along with data center resources. Data center operators are increasingly demanding a single unified mechanism to manage all the data centers in their domain instead of a separate, distributed management. SDN on the WAN is one of the mechanisms to achieve this.

4.2 SDN Controller

There are multiple SDN controllers in the industry today, supporting different deployment scenarios, each of which has pros and cons. The network controller provides a uniform and centralized programmatic interface to the entire network. It provides the ability to observe and control a network. On the northbound, it provides to the applications the network information base (NIB) with the network's observations such as nodes, links, topology, and statistics. Applications use this state to make management decisions. The controller software runs on commodity servers, which consists of multiple controller processes and provides a single consistent view.

Figure 4.5 shows an overview of the controller. On the southbound, controller installs instructions to switches. These forwarding

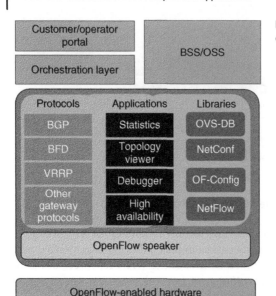

Figure 4.5 SDN controller overview.

instructions should be independent of the particular switch hardware and should support the flow-level control granularity. The controller provides high-level names and their bindings in the network view allows any application to convert a high-level name into low-level addresses. This high-level view is provided to the applications. Vertically, an SDN controller should contain three components: the protocol handler that deals with traditional network protocols, the set of applications that makes use of its network information, and the libraries that support various southbound interfaces. On top of the SDN controller, network orchestration tools, operational support system (OSS)/business support system (BSS), and other customized tools can be built.

When an incoming packet matches a flow entry at a switch, the switch updates the appropriate counters and applies the corresponding actions. If the packet does not match a flow entry, it is forwarded to a controller process. Controllers use these flow initiations and other forwarded traffic to construct the network view and determine where to forward. Besides packet-in event, the SDN controller also handles other network changes, such as link failure, node failure, and routing changes.

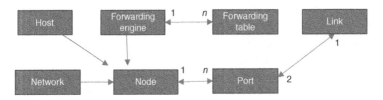

Figure 4.6 Example of SDN NIB.

One critical piece of all SDN controller is the NIB. It is a graph of all network entities, including switches, ports, interfaces, and links. The northbound applications read and manipulate the network through NIB. Each entity in the NIB is a key value pair. Examples of these network entities can be hosts, nodes, ports, network, and so on. NIB serves as a database to the applications. Thus, normal DB operations such as query, create, delete, access attributes, configuration, and pull are supported. Figure 4.6 shows an example of the NIB. Host, forwarding engine, and network are all nodes. Each node can contain multiple ports (shown as dashed line, and n means n ports). Each link has two ports.

4.2.1 Controller Deployment Choices

One of the main requirements for a centralized SDN controller to be deployed in a production network is its scalability and high availability. We propose that the key requirements for SDN scalability is that the centralized SDN controller can maintain high performance and availability with increasing network sizes, network events, and unexpected network failures. We focus our discussions on the controller scalability as follows.

To improve the controller scalability, a common practice is to deploy multiple controllers either for load balancing purposes or for backup purposes. According to different purposes, we can employ different models to provide a more scalable design. In the following, we discuss three different models for scalability and high availability for centralized SDN controllers: hot-standby model, distributed network information base, and a hybrid model. The last model incorporates important aspects of the previous two models. We argue the third model is more appropriate for complex network scenarios with carrier-grade scalability requirements.

Hot-standby model: This is similar to the current high availability solution for off-the-shelf router and switches. In this model, a master controller is protected by a hot-standby model. The standby instance will take over the network control upon failure of the master controller. The advantage of this model is its simplicity. However, it may encounter performance bottleneck when the number of switches and the communication messages grow significantly.

Distributed NIB: It employs a distributed system's concept, which is widely used in today's SDN solutions. In this model, the network controller is a cluster of controllers. Each of the controllers control a different part of the network. The network information is replicated across multiple controllers for high availability and scalability. This model is designed for large network with hundreds or even thousands of switches. The disadvantage is the communication overheads between controllers. With a careful design communication protocol, the drawbacks could be overcome. Some widely used SDN controller implementations, such as ONIX [80] or ONOS [81], belong to this model.

Hybrid model: This is a combination of the previous two models where the network information is replicated for high availability. In particular, controllers are grouped into the clusters. In each cluster, there is a master controller plus a hot-standby instance to handle the failure scenarios. It is organized in a hierarchical manner so that the scalability is guaranteed. This model is used in large-scale deployment, such as in AT&T's network.

One important question is how to distribute these data across multiple controller instance in a distributed system. Multiple techniques can be used to scale the controller. First, we can *partition* the data across multiple instances. Each instance can control a subset of switches and forwarding rules. Second, we can employ a *hierarchy* design. We can use zoom in and zoom out across different aggregation levels. Third, we can choose different *consistency* models for different data types. For example, network topology requires high accuracy at any given time, thus we should use strong consistency to it. But the network measurement data may not be as critical and can tolerate delay, so a weak consistency may suffice. For strong consistency, we can use traditional DB transactions. For weak consistency, we can use distributed hash table (DHT).

4.2.2 Apps on SDN Controller

The SDN architecture consists of application, control, and infrastructure layers. The interface between the application and control layers is how applications inform the network of their policy requirements. Examples of policy requirements are as follows: a security application may need all of an infected hosts a traffic redirected to a remediation server; a QoS application may need voice traffic delivered within a specified latency; and a troubleshooting application may need a copy of specific communication sent to a network administrator's protocol analyzer. Implementing their policies by writing low-level, prioritized flow table rules through a controller pass-through interface into the switches in the network has several problems.

- *Correctness*: When multiple SDN applications operate on the same network, the flow table rules written by these applications can conflict and cause the network and applications to fail to operate correctly. Since OpenFlow rules are prioritized, and only the first to match is executed, only one of the applications, the one that used the higher priority, will have their rules correctly enforced.
- *Coupling*: When an SDN application directly writes flow table rules, it must first comprehend the network design and infrastructure capabilities. In addition, an SDN controller may be using a mix of traditional protocols in a hybrid SDN network, along with OpenFlow. In this commonly occurring case, the application must understand the network design and technologies in use to successfully program rules into the network. If the controller is updated, or reconfigured, it may change the network design or set of technologies used and the application must also change its rules accordingly. This coupling between application and network design and infrastructure capabilities result in dependencies across application and controller product releases and version incompatibilities, complicating SDN network management by customers.
- *Ecosystem*: When SDN applications need to comprehend the network design and infrastructure capabilities, the applications are more complicated and the people required to write them must have knowledge of networking technology and operation. These factors make it hard to write SDN applications, resulting in slow

application development, sparse application offerings, a small SDN ecosystem, and limited value proposition for customers.

The fundamental problem facing such a deployment is that each app has its own objectives and may conflict with other apps over proposed changes to shared network infrastructure. These conflicts include competing for limited switch resources such as flow table entries, competing for network bandwidth, or powering on/off network devices. Addressing this problem poses a threefold challenge: (1) resource conflicts are dynamic and cannot always be determined a priori; (2) choice of how to resolve a conflict directly impacts global network objectives; (3) conflicts might actually be avoidable, and alternative allocations not initially expressed by apps may allow both intended requests to be satisfied.

There has been a lot of research in this space in SDN controller debugging, policy management, and conflict resolution. Statesman [82] automatically merges nonconflicting changes and relies on manually tuned priorities to resolve conflicts. OpenDaylight Network Intent Composition (NIC) [83] allow users and operators to describe high-level policy/intent rather than low-level device instructions. One recent approach is to explicitly model the policy of each SDN App and then use game theory approach to resolve conflicts [84]. It comprises a programming framework for SDN Apps and a configurable coordinator that detects and mediates these conflicts on behalf of the SDN controller.

4.3   SDN Data Plane

OpenFlow was initially proposed to decouple the routing intelligence (software) from simple forwarding (hardware) allowing, particularly for academic research networks and test beds, fast prototyping and evaluation of new control methods and algorithms.

OpenFlow provides an open control interface to the operating system of the network device without compromising the details of the implementation. This guarantees the business benefits for equipment manufacturers so that it will be more likely to be adopted by the industry. OpenFlow needs the support from the operating system. It is based on the ternary content addressable memory (TCAM)-based flow tables. In a classical router or switch, the fast packet-forwarding (data path) and the high-level routing decisions (control path) colocate on the same device. An OpenFlow-enabled switch separates these two functions. The data path portion still resides in the switch, but high-level routing decisions are moved to a flow controller, typically a standard server. The OpenFlow switch and controller communicate via the OpenFlow protocol, which defines operation and management (OAM) messages.

Figure 4.7 shows the basic component in the initial OpenFlow design. Later OpenFlow has evolved to other more complex structures such as multiple tables. The match field can match any packet headers, for example, Ethernet address, IP headers, or MPLS header fields. The typical actions can be forward, modify, and drop. Each entry is associated with a counter for collecting statistics purposes. Similar to routing or access control rules, each rule has a priority. When multiple rules match the same packet, the one with higher priority is selected. Finally, each rule has an expiration time. This is for preventing outdated rules. Once the time out is reached, the rule is removed by the switch.

With OpenFlow rules, each packet matches a specified header, the counters are updated, and the appropriate actions taken. If a packet matches multiple flow entries, the entry with the highest priority is chosen. An entry's header fields can contain wildcard values, meaning that it can match any value in the corresponding position. It is a TCAM-like match to flows. The basic set of OpenFlow actions are

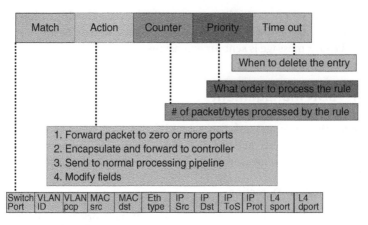

Figure 4.7 OpenFlow overview.

forward as default, forward out through a defined interface, deny, forward to a controller, and modify various packet header fields.

The messages can be initiated by the controller, by the switch, or by either of them.

- *Controller-to-Switch:*
  - The *Send-packet* message instructs the switch to send packet out of a specific port on a switch.
  - The *Flow-mod* message tells the switch to add/delete/modify flows in the flow table.
  - The *Read-state* message is used to collect statistics about flow table, ports, and individual flows.
  - *Features* is sent by controller when a switch connects to find out the features supported by a switch.
  - *Configuration* is to set and query configuration parameters in the switch.
- *Switch-to-Controller:*
  - The *Packet-in* message sends all packets that do not have a matching rule to controller.
  - The *Flow-removed* message is sent whenever a flow rule expires.
  - The *Port-status* is sent to the controller whenever a port configuration or state changes.
  - The *Error* is sent whenever an error occurs.
- *Both directions:*
  - *Hello* message is sent at connection startup stage to bootstrap the devices.
  - *Echo* is sent periodically to indicate latency, bandwidth, or liveliness of the connection between controller and switches.
  - *Vendor* message is sent for extensions.

4.4 SDN Management

Besides the basic network control and forwarding, SDN can enable new features in other network management area. In this section, we discuss three areas: measurement, failure recovery, and security.

4.4.1 Anomaly Detection

Network flow counting is essential to many network management applications, ranging from network planning, routing optimization, customer accounting, and anomaly detection. Today, statistics of traffic flows are reported by the routers to the centralized management system. Thus, the impact of network measurement on the network must be minimized. For example, an aggressive monitoring may result in artificial bottlenecks in the network. With too passive a scheme, it may miss important events. Thus, the key challenge is to strike a careful balance between effectiveness (supporting a wide range of applications with accurate statistics) and efficiency (intruding low overhead and cost). Among all the network management applications, from the security perspective, one important question to be answered is how to count flows to provide sufficient information for the network anomaly detection?

Existing attempts to achieve a better overhead/accuracy balance is through traffic sampling [85], that is, a router selectively records packets or flows randomly with a preconfigured sampling rate. The thinned traffic is then fed as input to anomaly detection. While being widely deployed as they are simple to implement with low CPU power and memory requirements, studies have shown it to be inaccurate, as it is likely to miss small flows entirely.

The network flow measurement should provide a more *flexible* and more *interactive* interface to the anomaly detectors so that the set of flows collected can be dynamically adjusted according to the findings of anomaly detector immediately. There are twofold benefits of such interfaces. On the one hand, the anomaly detector can instruct the flow collection module to provide finer-granularity data once there is a suspicion of attacks, so that the anomaly can be identified sooner. On the other hand, it can inform to collect coarser-grained flow data both spatially and temporally when there is no sign of attacks such that the traffic monitoring load is reduced. Therefore, this adaptive interface can simultaneously improve the accuracy and reduce the overhead.

SDN has two key features that enable the design of a flexible flow counting API for anomaly detection. On the one hand, SDN breaks the tight bindings between forwarding and counting. One can install separate wildcard rules on OpenFlow (OF) switches purely for monitoring purposes, which offers tremendous flexibility in defining the set of packets to count. On the other hand, thanks to the simple interface between control and forwarding plane, one can easily adjust the elements to count, by simply updating the counting rules. This property makes *real-time* adaptive counting possible.

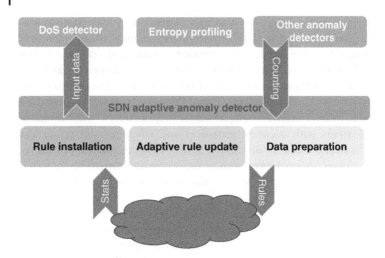

Figure 4.8 Adaptive SDN anomaly detection framework.

One idea proposed recently is to propose an adaptive flow collection method for anomaly detection in SDN [86]. Benefiting from the nature of SDN, it is programmable, allowing flexible specification of the spatial and temporal properties of counting units (or aggregates). Figure 4.8 shows the key idea of this approach. On the northbound, it provides input to different types of anomaly detectors such as DoS detector, traffic analysis using entropy profiling to detect abnormal distributions, and many other types. On the southbound, it instructs the switches to collect statistics in the entire flow space. It contains three components: the first and the most important component provides the adaptive zooming in the flow space and determines the set of rules to be installed. After that, the second component determines where the rules are to be installed in the network-wide in order to minimize the space taken on the switch flow tables. Finally, the third component prepares the data for anomaly detection applications, which will use these data. It anecdotes the data with the corresponding spatial/temporal aggregation information, which describes the accuracy of the data.

4.4.2 Network Measurement

In data center and Internet service provider (ISP) networks where operators own the entire network, monitoring on each device (i.e., a host or switch) should no longer be treated separately. To fully leverage the different views of traffic and different capabilities in monitoring different flow properties across devices, it becomes important to run coordinated measurements across different devices. For example, an ingress switch is a better place to measure per source traffic, while an egress is better at measuring per destination traffic. A host can monitor packet losses by collecting TCP-level statistics or packet-level traces, while switches can more easily measure the traffic through a network path.

Both hosts and switches have resource constraints when running these measurement tasks. Hosts need to devote most of their resources to the revenue-generating applications, leaving fewer processing resources for measurement. Switches have limited memory to store measurement data from all flows.

Temporal coordination of measurement across devices can significantly reduce measurement overhead. For example, we only need to start monitoring the per-flow volume at every hop of a path when we observe end-to-end anomalies at hosts (e.g., unexpected large flows, unexpected packet losses). Another example is diagnosing equal cost multi-path (ECMP) hashing problems. Rather than continuously monitoring all the flows at all the ECMP paths, we only need to start monitoring the flows whose traffic across any of the paths becomes large.

Given the resource limitations at individual hosts and switches, instead of monitoring all the flows all the time, we observe the benefits of coordinating these devices to monitor the right flows at the right time. We call this *temporal coordination* [87]. With temporal coordination, we can leverage limited resources at hosts and switches to only capture the important flows of interest. For this, we need *monitors* that decide when and which flows to monitor and *watchers* that collect information for the selected flows at the right time. We now give a few examples highlighting the benefits of temporal coordination.

We discuss one potential solution of the temporal correlation that can be enabled by SDN as follows. Figure 4.9 describes one such temporal correlation measurement system. It configures two key types of components for each monitoring task: the *monitors* that capture network events, select related flows, and send the information to the *watchers* that will collect flow-level statistics for the selected flows. To ensure high monitoring accuracy and reduce the memory usage, it is important to coordinate the monitors and watchers in a timely fashion. We can develop coordination algorithms that allow

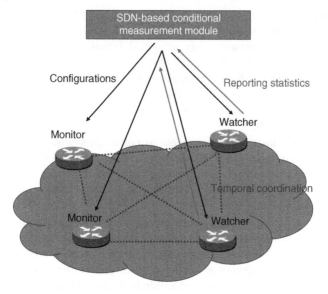

Figure 4.9 Conditional SDN measurement overview.

the monitors to make the best use of their limited memory based on the information from the selectors. Since the monitors and watchers may be located in different devices, we introduce packet tagging and explicit messaging to ensure the efficient communication between the monitors and watchers ensuring high monitoring accuracy with low bandwidth overhead. To support the maximum number of monitoring tasks in a network of devices with limited memory, we can use optimization techniques to identify the best locations of monitors and watchers by considering the relation between monitors and watchers and the latency constraint between them.

4.4.3 Failure Recovery

The susceptibility to failures is always a key factor for the prosperity of any network technologies. In the past, despite that many efforts have been put to add reliability and high availability to networking systems, its performance can still be severely impacted by hardware failures, software bugs, and configuration errors. Parallel to the efforts in developing new debugging tools, another approach is to improve on the support for recovery from errors, assuming that errors are inevitable.

Checkpointing is a common and powerful approach to recover from transient errors in servers and distributed systems. In general, a system periodically records its state during normal operation and stores it in nonvolatile storage that is, checkpointing. Upon failure, it is restored to a previous state and restarts the execution from this intermediate state, that is, rollback process, thereby reducing the lost computation. This technique is especially useful for long-running applications such as scientific computing and telecom applications, where restarting from the beginning can be costly. Besides host-level process migration and debugging, it can be used for error recovery in distributed systems, where a collection of application processes are checkpointed distributively, and a globally consistent state can be restored. Checkpoints can also be used for debugging and root cause analysis.

Despite its usefulness, checkpointing and rollback are rarely used in networking infrastructure for the following reasons. First, the network interacts with the outside world constantly, that is, receiving packets and sending them out, the outside world cannot be rolled back. Second, traditional checkpoint-recovery mechanism assumes a fail-stop system, that is, upon any fault or failure, the process or system terminates. This assumption does not always hold true in networking. For example, a loop may exist for a nonnegligible amount of time before it is detected, even if it is hurting the performance. Third, the network equipments and applications are mostly as black boxes to the operators, making it impossible to instrument and reason. Fourth, the network state can be very huge, making the storage of checkpoints costly. Thus, in traditional network, the fault recovery process is still error-prone and notoriously hard.

SDN may make checkpointing and rollback feasible, thanks to its three key properties: simple abstraction, network wide visibility, and direct control. The network can be viewed as a distributed collection of switches managed by a logically centralized control programs with a global network view. The controller reads and writes directly into flow tables on each switch in the form of rules through a standard API OpenFlow. Each rule contains a matching field and a set of actions. Exploiting these properties of SDN, in this work, we explore a *scalable way* to perform checkpointing and rollback framework in networks for fault recovery.

To recover from fault in a network, three principal steps are involved: detecting a fault, containing it, and rolling the nodes of the network back to a certain checkpoint. SDN can be used to allow the entire network state to roll back to a consistent global state, to mitigate the impact of faults. The network state includes both the controller states and the forwarding tables in all the switches in the network. To enable fast and independent recovery from switches, we support that each switch independently performs checkpoint.

Figure 4.10 shows the solution overview of this approach. First, checkpointing the switch means taking snapshots of the flow tables in each switch. Since the rules are installed by the controller, the straightforward way is to let the controller keep track of the set of rules installed on each switch and then consider it as a part of the controller state. However, somewhat surprisingly, we found that most of the existing controllers maintain a copy of the flow tables for the switches, mainly because of the concerns of space and overhead. Instead, we can take a distributed approach. Each switch is responsible for checkpointing its own state. For rollback recovery, some coordination between the controller and the switches can be used to select a network-wide consistent set of previously collected checkpoints and roll each switch back to the checkpoint selected for that switch. Second, controller checkpoint is similar to checkpoint of any type of user-level process. That is, a checkpoint of the controller includes the controller process's address space and the state of its registers. For recovery, the new process is spawned, which initializes its address space from the checkpoint file and resets its registers. It is achieved by adding a piece of code, called checkpointer, being compiled or linked with the controller transparently.

4.4.4 Controller Placement

In evaluating a network design, the network resilience is an important factor, as a failure of a few milliseconds may easily result in terabyte data losses on high-speed links. In traditional networks, where both control and data packets are transmitted on the same link, the control and data information are equally affected when a failure happens. The existing work on the network resilience analysis have therefore assumed an in-band control model, meaning that the control plane and data plane have the same resilience properties. However, this model is not applicable to split-architecture networks. On one hand, the control packets in split-architecture networks can be transmitted on different

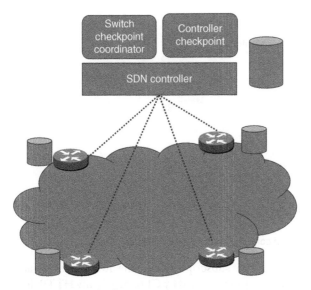

Figure 4.10 SDN checkpoint system.

paths from the data packet (or even on a separate network). Therefore, the reliability of the control plane in these networks is no longer linked with that of the forwarding plane. On the other hand, disconnection between controller and the forwarding planes in the split architecture could disable the forwarding plane: when a switch is disconnected from its control plane, it cannot receive any instructions on how to forward new flows and becomes practically offline.

In the following, we illustrate the reliability of SDN in an example in Figure 4.11, which consists of seven OpenFlow switches and two controllers. For simplicity of illustration, we assume fixed binding between controller and switches, which is the shortest path between the switch and its closest controller. Another assumption is the static binding between controller and the switch, for example, C1 is the assigned controller for S3. S3 can only be controlled by C1 even if it is also reachable by C2. In this example, we assume that there is a separate link between two controllers C1 and C2 to exchange the network states between them. Each controller uses the same infrastructure (network) to reach the OpenFlow switches. For instance, S7 goes through S3 and S1 to reach the controller C1, marked as dotted line. We also assume fixed

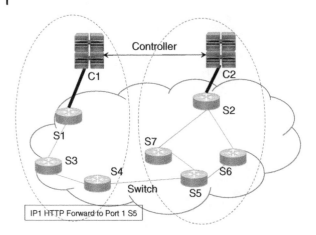

Figure 4.11 Example of controller and switch connection.

routing has been set up. The subscripts denote the flow entries in each switch. An entry on S4 is programmed by C1 to match any HTTP flow from IP1 and forward to port 1 connected to S7.

If the link between S4 and S5 fails, connections between any of the switches S1, S3, or S4 to any of the switches S2, S5, S6, or S7 would be interrupted. If the link between S1 and controller C1 fails, then until a backup path is built and used, S1 will lose its connection to its controller. Assuming that in this case the switch invalidates all its entries, and then S1 cannot reach any other switch in the network, until it reconnects to its controller. This is like S1 itself is failed for a period of time.

The types of failures in SDN can be categorized into three types:

- *Link failure*: A link failure indicates that traffic traversing the link can no longer be transferred over the link. The failure can be either of a link between two switches or of a link between one controller and the switch it connects to. We assume network links fail independently.

- *Switch failure*: A switch failure indicates that the corresponding node is unable to originate, respond, or forward any packet. Switch failures can be caused by software bugs, hardware failures, misconfigurations, and so on. Again, we assume network nodes fail independently.

- *Special case – connectivity loss between switch and controller*: A switch may lose connectivity to its controller due to failures on the intermediate links or nodes along the path. In this invention, we assume that whenever a switch cannot reach its controller, the switch will discard all the packets on the forwarding plane, even though the path on the forwarding plane is still valid. Therefore, this can be considered as a special case of switch failure.

4.4.4.1 A Special Study: Controller to Switch Connectivity

In the following, we illustrate the reliability issue using a specific problem. We focus on the controller placement problem given the distribution of forwarding plane switches. We consider that control platform consists of a set of commodity servers connecting to one or more of the switches. Therefore, the control plane and data plane are in the same network domain.

The connectivity to the controller is extremely important for the OpenFlow network reliability. We define the reliability of controller to data plane as the average likelihood of loss of connectivity between the controller and any of the OpenFlow switches.

In the following, we discuss three aspects of the connectivity between controller and the switches.

Routing between controller and switches

For a given controller location, the controller can construct any desired routing tree, for example, a routing tree that maximizes the protection of the network against component failures or a routing tree that optimizes the performance based on any desired metrics. One of the popular routing method is the shortest path routing constructed by intra-domain routing protocols such as open shortest path first (OSPF). The main problem with the shortest-path routing policy is that it does not consider the network resilience (protection) factor. To maximize the routing, one can develop an algorithm with the objective of constructing a shortest-path tree. Among all possible shortest-path trees, we can find the one that results in best resilience compared to other shortest-path trees.

The effect of protection depends on both on the selection of the primary paths and the choice of the controller location. To get a sense of how these two factors affect the protection metrics in an SDN, we make some calculations in two example networks, with the topology of Internet2 [88] network in Figure 4.12 and a typical Fat-tree like data center network in Figure 4.13.

Figure 4.12 shows the Internet2 topology with 10 nodes and 13 edges. Here we illustrate two examples of controller selection,

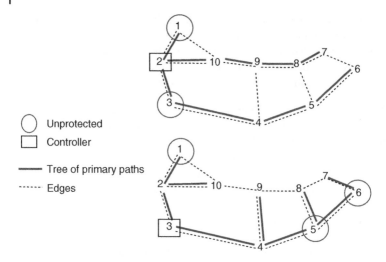

Figure 4.12 Example of Internet2 network protection.

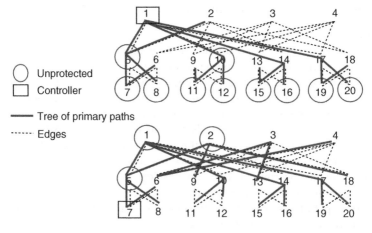

Figure 4.13 Example of Fat-tree network protection.

deploying controller on node 2 and node 3. The location of the controller is shown in rectangular box. For each controller deployment, we compute the shortest path tree to all other nodes (switches), rooted in the controller. The primary paths in the tree are shown in solid lines. For each switch, we compute its backup path. The nodes without any protection are shown in circles. In this example, we observe that the upper case is a better solution, since only two nodes are unprotected. In particular, node 5 is protected by node 8 in the upper figure but not in the bottom figure since it is the parent of node 8 in the primary path.

OpenFlow type of SDN also raises a large amount of interests in enterprise and data center networks. Figure 4.13 shows another example in Fat-tree like data center networks, containing 20 nodes and 32 links. Similarly, we illustrate two scenarios, deploying controller on node 1 and node 7. We construct the two shortest path trees rooted from the controller. Interestingly, we found that node 7 is a much better location for controller compared to node 1. This is very counterintuitive, as node 1 as the core nodes in the Fat-tree structure were expected to be more resilient. When node 1 is the controller, 10 nodes are unprotected, including all the leaf nodes. With node 7 as the controller, only three nodes are unprotected. For example, node 11 is protected by node 9 when its parent in the primary path (node 10) fails.

From these two examples, we observe that first, the controller's location does have large impact on the number of protected nodes in the network and second, the selection of controller's location can be quite counterintuitive. These observations motivate us to investigate a systematic approach of controller placement to maximize the protection in the split architecture network.

Deploying multiple controllers

Next, we consider the problem of deploying multiple controllers in the network. The problem can be formulated as following. Given a network graph, with node representing network's switches and edge representing network's links (which are assumed to be bidirectional), the objective is to pick a subset of the nodes, among all candidate nodes and colocate controllers with switches in these nodes so that the total failure likelihood is minimized. Once these nodes are selected, a solution to assign switches to controllers is also needed to achieve maximum resilience. The problem can be solved as a graph partitioning or clustering problem. The details are described in [89]. Two algorithms are proposed for SDN with multiple controllers. It is shown that the choices of controller locations do have significant impact on the entire SDN reliability.

In SDN, the best resilient scenario is that each switch has a preconfigured backup path to the controller. The backup path can be used to reconnect to the controller if any failure is detected on the primary path. Such rerouting behavior requires no intervention from the controller and no changes to the other switches in the network. In the following, we describe our design choices and the protection metrics in detail.

With the high requirements on network reliability, we need a mechanism to improve the resilience of the connectivity between the controller and the switches in a split architecture network. The mechanism should meet the following requirements:

- The protection should resume the forwarding after a failure as soon as possible. The existing IGP protocols such as OSPF and IS-IS typically take several seconds to converge, which cannot meet the sub-50 ms level of failure recovery time. One option is to rely on the controller to detect the failures in switches or links using some implicit mechanisms, for example, when hello messages are not received by the controller from a switch. This method will introduce a large delay in the network for failure detection and service restoration.
- The decision of protection should be made locally and independently. Different from traditional network, the forwarding element does not have a complete topology of the network. It only receives forwarding rules from the controller. When losing the connectivity to the controller, the switch has to make the decision of failover independently without any instructions from the controller. In other words, there will only be a local change in the outgoing interface of the affected switch. All other connections in the network will remain intact.
- We should keep the forwarding element, that is, the switch, as simple as possible.

We denote a network of switches by a graph $G = (V, E)$, where V is the set of nodes (switches) in the network and E the set of bidirectional edges (links) between nodes. With this given topology, we want to know at which node the controller should be deployed.

Our first assumption is that the network controller is in the same physical network as the switches. That is, we want to use the existing infrastructure of the network (i.e., existing links and switches) to connect the controller to all the switches in the network, as opposed to using a separate infrastructure (by adding new links and switches in the network) to connect the controller to the switches.

A cost is associated with each link based on which the shortest path routes between any two nodes are calculated. We assume that the cost on each link applies to both directions of the link.

We assume that there is no load balancing on the control traffic sent between the switches and the controller. Therefore, each node has only one path to reach the controller. In other words, the control traffic is sent from/to the controller over a tree, rooted at the controller, which we will refer to by *controller routing tree*. This routing tree covers all the nodes in the network and a subset of the edges. We assume that the same routing tree will be used for communications between the controller and the switches in both directions.

With a given controller location, different routing mechanisms could be used to form different routing trees. Figure 4.14 shows a network and its controller routing tree. In this figure, the dashed lines show all links in the network, and the solid lines show the links used in the controller routing tree. Each node can reach the controller by sending its control traffic along the paths in the controller routing tree. We assume that both directions of each link have the same cost and therefore, they are symmetric.

In the controller routing tree T, node u is an upstream node of node v if there is a path in T from node v to node u toward the

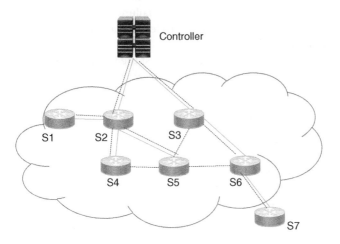

Figure 4.14 Protection against link and node failures.

controller. Node u is called a downstream node of node v if there is a path in T from node u to node v toward the controller. In the network depicted in Figure 4.14, for example, node S_3 is an upstream node of nodes S_6 and S_7, and these two nodes are downstream nodes of node S_3. In the controller routing tree, a node's parent is its immediate upstream node and a node's children are its immediate downstream nodes. Because of the assumed tree structure, each node has only one immediate upstream node in T.

4.4.4.2 Node Versus Link Failure

With a given controller location and controller routing tree T, consider node a and its immediate upstream node b. We say node a is protected against the failure of its outgoing link (a, b) if there exists node $c \in V\backslash\{a, b\}$ with the following properties:

1) (a, c) is in G (i.e., there is a link between nodes a and c in the network).
2) Node c is not a downstream node of node a in T.

The second condition guarantees that a loop will not be created as a result of connecting node a to node c.

If the above conditions are met, then link (a, c) could be assigned as the backup link for link (a, b), and this backup link could be preconfigured in node a. As soon as node a detects a failure in link (a, b), it will immediately change its route to the controller by changing the primary link (a, b) to the secondary link (a, c).

We say that node a is also protected against the failure of its immediate upstream node if node c satisfies another condition in addition to the above ones.

3) Node c is not a downstream node of node b in T.

This condition guarantees that the control traffic of node c toward the controller will not pass through node b (which is assumed to have failed). Again, as soon as node a detects a failure in node b, it switches its outgoing link over from (a, b) to (a, c).

In the network shown in Figure 4.14, for example,

- Switches S_1, S_2, and S_7, are not locally protected, that is, if their outgoing links (or upstream nodes) fail, no backup links can be chosen and preconfigured to be used.

- Switch S_4 is protected against its output link failure, that is, if link (S_4, S_2) fails, link (S_4, S_5) could be used instead. However, S_4 is not protected against its immediate upstream node (S_2) failure. Because the backup path $(S_4, S_5, S_2, \text{controller})$ will pass through S_2.
- Switch S_6 is protected against both its outgoing link and its immediate upstream node failures: If link (S_6, S_3) or node S_3 fails, the control traffic of S_3 will be sent over link (S_6, S_5) and it will not pass through node S_3.

Depending on how critical or frequent link failures are versus node failures in the network, the network operator could assign different costs to these two kinds of failures, for example, cost α for node failure and cost β for link failure. For example, $\alpha = \beta$ could be interpreted and used for scenarios where link and node failures are equally likely – or when it is equally important to protect the network against both kinds of failures. This way, the cost of not having protection at a node could be evaluated at $\alpha + \beta$ if the node is not protected at all, at α if it is protected only against its outgoing link failure, and at zero if it is protected against the upstream node failure as well. For those switches directly connected to the controller, the upstream node protection cannot be defined (as the immediate upstream node is the controller). For those nodes, therefore, the assigned cost is zero if they are protected against their outgoing link failure and is $\alpha + \beta$ otherwise.

4.4.4.3 Downstream Versus Upstream Nodes

In this work, we assume that the traditional failure management tools are deployed in the split-architecture network, that is, there is no extended signaling mechanism for a node to inform its downstream nodes of a failure. Therefore, if a switch is disconnected from the controller (i.e., if there is no backup path programmed in the switch), then all its downstream nodes will also be disconnected, even if they are themselves locally protected against their outgoing links or immediate upstream nodes failures. This means that in evaluating networks resiliency, more weights should be assigned to nodes closer to the controller (which is the root of the controller routing tree). More precisely, the weight of each node should also be proportional to the number of its downstream nodes.

In Figure 4.14, for example, failure of the link between S_2 and the controller results in the disconnection of all S_1, S_2, S_4, and S_5 from the controller. This failure costs four times more compared to when the link (S_1, S_2) fails, which only disconnects S_1 from the controller.

Implementation of protection using OpenFlow

Our proposal can be applied to any implementation of the split architecture. The forwarding table in an OpenFlow switch, for example, is populated with entries consisting of a rule defining matches for fields in packet headers, a set of actions associated with the flow match, and a collection of statistics on the flow. The OpenFlow specification version 1.1 introduces a method for allowing a single flow-match trigger forwarding on more than one port of the switch. Fast failover is one of such methods. Using this method, the switch executes the first, that is, live action set. Each action set is associated with a special port that controls its liveliness. OpenFlow's fast failover method enables the switch to change forwarding without requiring a round trip to the controller.

4.5 SDN Security Attack Prevention

Network attacks have long been an important problem and have attracted a lot of research in academic and commercial sector. With a rapidly growing number of critical as well as business applications deployed on the Internet today, network attacks have both become more lucrative for the attackers and more damaging to the victims. The implications of network attacks on the victim can be huge. For example, a distributed denial-of-service (DDoS) can overwhelm the victim and make it unable to handle its regular business. A large-volume DDoS attack can further cause collateral damage to traffic that shares links with the victim's traffic, leading to large traffic drops, BGP session interruptions, and routing interruptions [90]. Besides the data plane attacks, control plane misconfigurations and attacks on the interdomain routing protocol BGP can have dire implications for victim networks. For example, the prefix-hijacking attack injects and propagates false routes to the Internet, causing victim's traffic to be redirected to the attacker networks for sniffing, modification, or dropping [91]. Traffic sniffing and modification are very difficult to detect and mitigate, and create huge security and privacy issues for the victim, while blackholing severely affects online businesses and critical infrastructures.

Many solutions have been proposed to detect and mitigate *individual* attacks. For example, in DDoS realm, many victim-deployed or ISP-deployed DDoS defenses, overlay-based DDoS defenses [92], and content replication to sustain high-volume attacks have been proposed and deployed. In routing realm, detection approaches that monitor live BGP data feeds and conduct data plane probing have been proposed to diagnose prefix-hijacking attacks.

But ultimately, traffic flows, attacks, and their routes are the results of actions of multiple networks, each following its individual interests and priorities. Thus, while many attack instances can be handled by the victim and its local ISP, there will always exist attacks that cannot be diagnosed or mitigated without help from remote networks, which are involved in sourcing or carrying traffic to the victim. Today's Internet lacks such wide-scale, general service for automated inter-ISP collaboration on security problem diagnosis and mitigation.

There have been numerous research works on inter-ISP collaboration for attack diagnosis and mitigation, such as collaborative DDoS defenses, (Defense-by-Offense, Internet traceback [93], pushback [94], DefCOM [95]), collaborative spoofing defenses (Packet Passports [96], Hash-based traceback [93], core-based traffic filtering) collaborative worm defenses, and collaborative routing defenses. However, most proposals are still not deployed today because (1) most of the proposals only focus on detection or mitigation of one attack type or variant; (2) some solutions require complex changes of the data plane or new router functionality, which are difficult to achieve; and (3) some solutions do not create proper incentives for ISPs to collaborate with each other.

Inspired by SDN, which provides a simple interface (flow-based rules) to facilitate the advancement of networking protocols, we propose Software dEfiNed Security Service (SENSS), a generic interface for Internet attack diagnosis and mitigation. SENSS has three key features as follows:

1) *Victim-oriented:* The victim of a security attack has the most incentive to detect and mitigate the attacks. The victim also has the most knowledge about the problems it is experiencing, the traffic it sees, and the help it needs to diagnose and remedy problems. Our proposed service enables this victim to directly request security services from multiple remote ISPs. For security and privacy reasons, we design mechanisms for victims to only have visibility

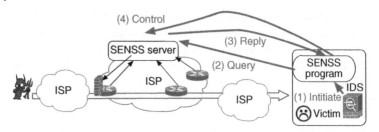

Figure 4.15 SENSS architecture.

and control of their own traffic, that is, the traffic that carries either source or destination IPs from the victim's address space.

2) *Simple detection/mitigation interface from an ISP:* We define a simple interface for victims to request services from ISPs, such as statistics gathering, traffic filtering, rerouting, or QoS guarantees. The interface is both expressive to support the detection/ mitigation of a variety of attacks (e.g., prefix hijacking, DDoS), and easy to implement in today's ISPs.

3) *Programmable attack detection and mitigation across ISPs:* With the simple interface provided by ISPs, victims can easily *program* their own attack detection and mitigation solutions across ASes. As shown in Figure 4.15, a victim can first query multiple ISPs to trace back the attack, identify the best locations for remediation, and then issue commands for ISPs to take mitigation actions (e.g., filtering the traffic, guaranteeing bandwidth or rerouting).

End networks have every incentive to use SENSS, since they receive much needed help and they pay only when they use the service. SENSS is economically appealing for ISPs because they can charge the victims for the services they provide.

4.6   SDN Traffic Engineering

To optimize network costs, performance, and throughput, network operators need to carefully engineer their network traffic. SDN enhances traffic engineering (TE) by offering programmability and global control over the forwarding plane. Recently, service providers have designed centralized TE applications on the SDN controller (e.g., Google's B4 [97] and BwE [98]) to globally compute and enforce optimal paths and bandwidth allocations for all flows in their network.

At first sight, it seems these solutions can be adapted to manage traffic in ISP networks. However, a closer inspection reveals they fall short in satisfying scalability requirements of ISP networks.

These SDN-driven TE systems are designed to manage data center (DC) interconnects with at most tens of nodes (e.g., Microsoft's SWAN [99]). However, large ISP networks can consist of tens of thousands of forwarding devices. Therefore, global flow optimizations, which are built based on linear programming (LP), can have millions of constraints and variables. This easily makes the existing centralized TE intractable because of its exponential worst-case complexity [100]. Adding more compute resources to the TE application does not fundamentally address the scalability challenge given the rapid growth of the network size. Having multiple SDN controllers (e.g., ONOS [81]) does not by itself reduce the TE complexity because the flow optimization finally is assigned to a single instance, needing to process the entire TE states (e.g., topology, traffic matrices, constraints).

To address these issues, we present a software-defined TE solution, called SdnTE, that can scale to manage traffic in very ISP networks. SdnTE enables ISPs to execute TE rapidly and frequently using a logically centralized software in their network. At its heart, SdnTE sacrifices a small amount of network throughput compared to the global, optimal TE techniques to substantially accelerate the TE computation by several orders of magnitude (e.g., from hours to seconds). SdnTE enables ISPs to quickly respond to unexpected changes in traffic patterns (e.g., unpredictable rush hours), and thus highly utilize their network resources. In more detail, SdnTE builds up a hierarchical control plane for ISPs and designs a novel recursive, distributed TE application on it. For TE scalability, SdnTE systematically partitions the global TE task throughout hierarchy by creating a logical TE region for each control node in the hierarchy. Each node is responsible for optimizing flows in its region using linear optimization, while coordinating with its parent and children indirectly.

Historically, the recursive topology aggregation and hierarchy techniques have been used to minimize the routing table size in routing protocols (e.g., PNNI [101], Nimrod [102]) in different network technologies. These designs are limited to the distributed link-state or source-based routing protocols and cannot be leveraged for global flow optimization. Hierarchical SDN controllers have been used to scale other network applications (e.g., mobility [103]), not on the centralized TE problem. In the presence of recursive abstractions, it is not straightforward to have the LP-based hierarchical TE generate feasible results, and efficiently and highly utilize network resources. For the same reason, enforcing TE results with minimal forwarding states in the data plane is challenging.

4.6.1 TE Architecture and Solution Overview

Consider a software-defined ISP network consisting of thousands of programmables switches distributed throughout a large geographical area (continent or country). In this network, many flows are generated between access switches (from home routers or peering points). Considering the ISP size, a global, optimal flow optimization can be very slow in practice (easily taking hours) and can lead to tremendous forwarding states. Our goal is to design a scalable control and data plane TE solution for the ISP networks that can *simultaneously* reduce the TE time (to orders of seconds) and shrink forwarding tables. We begin with an overview on the architecture, its design challenges, and our solutions.

4.6.1.1 Hierarchical Control Structure

For TE scalability, we build a hierarchical control logic for the network to break the global TE problem into smaller pieces. As shown in Figure 4.16a, our TE control logic consists of a set of TE control nodes (*cNode*) organized in a logical tree structure. The root cNode (e.g., cN7) is at the top and leaf cNodes (e.g., cN1–4) are at the bottom. Each cNode is a piece of software and solves a small part of the global TE problem. To benefit from locality, cNodes can be distributed among PoPs close to the switches. To systematically distribute the TE load and states (i.e., variables, constraints) among cNodes, we recursively and automatically construct an abstract region for each of them: the TE framework partitions the physical data plane (level 1) into a set of logical TE regions (e.g., regions 1–4) based on factors defined by the ISP (e.g., traffic patterns). Each leaf region is assigned to a leaf cNode that constructs an abstract view of its region and then presents it to its parent. Then each level-2 cNode (e.g., cN5–6) fetches its logical region topology from the children, and then abstracts it for its parent. Recursively, the process creates *increasingly more simplified* topologies until the root cNode (e.g., CN7) forms its abstract region. In this process, each cNode hides the detailed topology of its region and constructs a single switch for its parent called SdnTE Switch (hSwitch). A hSwitch

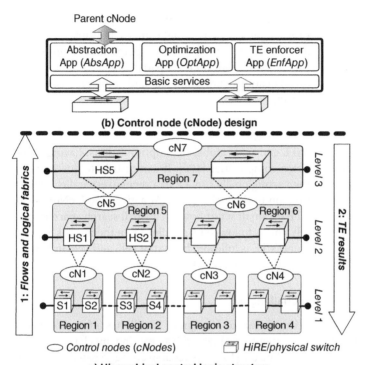

Figure 4.16 SdnTE network architecture.

is a *new* software abstraction over a physical/logical switch topology. In Figure 4.16, switches in region 2 are abstracted as hSwitch HS2 in region 5. hSwitch ports are logical, each corresponding to one or more border switch ports in the constituent physical topology.

4.6.1.2   Control Node Design

The software architecture of cNodes are illustrated in Figure 4.16 that consists of three applications: *Abstraction App (AbsApp), Flow optimization App (OptApp)*, and *TE enforcer App (EnfApp)*. AbsApp is responsible for communicating with the parent cNode by building and through the abstract hSwitch. TEOpt makes local TE decisions for flows appearing in its region. Although SdnTE is flexible to support different TE techniques (e.g., greedy, linear, and nonlinear), we focus on a designing multipath flow optimization using LP: TEOpt iteratively collects flow requests in the next interval from its region (*i.e.*, set ⟨source, sink, volume⟩s), and takes constraints, and objective function as other inputs. Then, it determines multiple paths for each flow (e.g., *k*-shortest paths) and computes locally optimized bandwidth allocations to them on their paths. The TE results (paths and bandwidth allocations) are enforced and implemented in the region using EnfApp.

4.6.1.3 Scalability Benefit

Intuitively, SdnTE can scale global flow optimizations, having exponential worst-case complexities [100]. It can distribute flows among regions, and then semiglobally run TE in a recursive-parallel manner from the root region to the leaf regions. At a given level, cNodes can perform their TE on small set of flows in a small search space in parallel. Recursively, this is followed by parallel TE in the level as follows. In Figure 4.16, cN7's TE is followed by that of cN5 and cN6 in parallel and then parallelism among cN1–4.

4.6.2 Design Challenges

A close inspection reveals our design of SdnTE has a few challenges. In the literature, there are hierarchical routing mechanisms (e.g., PNNI [101] and Nimrod [102]) in recursively aggregated networks. However, these systems do not provide solutions to our design challenges because SdnTE is a centralized, multipath *TE solution* in the SDN environment. In contrast, they are distributed, single-path *routing* protocols in nonprogrammable networks. These systems do not optimize flows, mostly deal with the path computation, and at best are equipped with simple circuit reservation mechanisms. Compared to them, SdnTE is challenging as it needs to semiglobally plan flows based on the ISP-selected objective functions such that network resources are highly utilized. In particular, these routing systems are inefficient from TE aspects. In general, poor network utilization, circuit reservation failures, and congestion are inherent to them. We now elaborate on the SdnTE challenges.

Challenge #1 – optimized TE in the presence of recursive abstractions: In the flat SDN, a single TE application has full control on switches, flows, and traffic; it can compute globally optimized paths and bandwidth allocations by running linear flow optimizations (e.g., [99, 104]). For TE scalability, SdnTE exposes a partial and abstract region to each cNode and distribute flows among regions. The design of SdnTE can easily lead to inefficient and infeasible paths and bandwidth allocations when TE is recursively computed from the root.

Recursively, when a cNode learns its parent flows engineered over the abstract hSwitch, it locally makes TE decisions to distribute them in its region.

Challenge #2 – efficient TE enforcement without direct data plane control: In the flat SDN, the TE application can easily enforce the TE results by establishing tunnels for each flow between its endpoints. In SdnTE, there is no cNode with full visibility and control on all switches. In fact, multiple cNodes might make partial TE decisions on flows recursively. Mappings partial TE results in regions to the physical data plane without full topology visibility is not only challenging but also can cause tremendous forwarding states (switch rules or packet instructions) in the physical data plane.

In SDN, the TE application has full control over the data plane switches, it can implement TE results by programming switches with tunnels (e.g., MPLS tunnels) for scalability (e.g., [97, 99]). In SdnTE, when the TE is computed from the root, multiple cNodes might make partial TE decisions on the same flow from its source to sink. Each cNode splits it into multiple tunnels, allocates bandwidth on them, and then programs its local hSwitches. In SdnTE, scalable mappings of partial tunnels in logical regions to the physical data plane without global topology visibility is challenging. Unfortunately, existing solutions either lead to high header overhead and throughput loss (e.g., label stacking in PNNI [101] and HSDN [105]), or result in the flow table explosion in SDN switches (e.g., label swapping in SoftMoW [103]).

4.6.3 TE Solution Overview

In response to the above issues, we enhance SdnTE with two set of procedures propagating necessary information between hierarchical regions. In each TE epoch (iteration), they run in a recursive fashion in the hierarchy. We first provide an overview and delve into the details in the later sections.

- *Bottom-up recursion:* In SdnTE, each TE epoch starts with smart assignment of flow requests to cNodes to efficiently distribute the

TE load. This is accompanied by providing some hints to cNodes regarding the physical topology to allow them efficiently participate in the global TE. Starting from the leaf level, recursively upward, each cNode (through AbsApp) collects flow requests from its region. Then, it makes local decision on whether or not offload some of the flows its parent region through the abstract hSwitch. Also, it associates a logical fabric to the abstract hSwitch and sends it to the parent. A fabric is a compact, dynamic representation of performance and resources in the cNode's region. However, it is more than a fixed and lossy graph built from the physical topology (e.g., PNNI [101]). In fact, it is computed based on the cNode's local flows and objective function as well as its parent's behavior in the past. The bottom-up procedure stops when the root cNode learns its flows and fabrics.

- *Top-down recursion*: In SdnTE, each TE epoch finishes with computing TE results and enforcing them in the physical data plane. Recursively downward, each cNode through AbsApp (except root) receives some flow requests on the abstract hSwitch from its parent when after the TE in the upper-level region. Considering the local hSwitch's fabrics, each cNode's OptApp computes multiple paths for each of its local and parent flows on its abstract topology. Then it allocates bandwidth to them by running a linear program. Finally, each cNode's EnfApp programs its region to split flows based on the TE results into multiple label-based tunnels (e.g., MPLS) between its endpoints. Each cNode locally minimizes the forwarding states that needs to be pushed into its local hSwitchs or child regions. For this purpose, EnfApp intelligently swaps and stacks tunnel labels in the packets. The top-down procedure finishes when leaf cNodes run their TE and program the physical switches.

4.7 Summary

SDN architecture introduces a separation between the control and forwarding components of the network. Among the use cases of such architecture are the access/aggregation domain of carrier-grade networks, mobile backhaul, data center networks, and cloud infrastructure, all of which are among the main building blocks of today's network infrastructure. Therefore, proper design, management, and performance optimization of these networks are of great importance.

In this chapter, we first provide an overview of the design principles and building blocks of the SDN architecture. We discuss the use cases, the challenges, and its business impact. Then, we dive into more details of SDN controller and SDN data plane separately. Finally, we conclude with introducing a few interesting and novel applications that SDN enables.

5

SDN and NFV in 5G

Mobile network operators are looking for cheaper and more efficient ways to connect subscribers/devices to their networks by searching for creative ways to minimize network traffic and latency. Next-generation 5G mobile networks will steer mobile network operators toward software-defined networks (SDN), agnostic network access, mobile edge computing, and 5G network slicing where grouped subscribers or machine-to-Machine (M2M) and Internet of things (IoT) devices are serviced by separate, virtualized core networks. Next-generation 5G mobile networks will merge IT and Cloud concepts into mobile core networks methods for accessing subscriber information in order to reduce data latency and network backhaul. Mobile edge computing will position network applications, service applications, and subscriber profiles on the network edge in close proximity to the subscriber/device in order for 5G networks to accomplish their goals, thus making services and profile as mobile as the subscriber and device.

In a 5G network, the edge locality is called an anchor point where network access and service processing is performed in an evolved packet core (EPC) network deployed on an SDN. An anchor point can be defined for a specific network slice of common subscriber's devices. Anchor switches provide network traffic routing between SDN anchor points. In contrast to 5G networks, 2G/3G/4G networks are generally organized as geographical regions serviced out of a regional data center. This regional data center may use one or more EPC SDNs similar to a 5G anchor point to service subscribers/devices in that region.

Network Function Virtualization: Concepts and Applicability in 5G Networks, First Edition. Ying Zhang.
© 2018 John Wiley & Sons, Inc. Published 2018 by John Wiley & Sons, Inc.

In this chapter, we first present the 5G overview and its relation with SDN/NFV. Then, we discuss service function chaining (SFC) in detail, which is a key concept in 5G deployment, and later present the virtualization usage in 5G: virtualized EPC.

5.1 5G Overview

As the next-generation mobile network, 5G is proposed when the number of user equipment (UE) and the surge in bandwidth has significantly increased. There are several objectives that 5G network is designed for, as follows:

- *Supporting massive amount of connected devices*: With the deployment and evolution of IoT, it is predicted that by 2010, there would be 50 billion Internet-connected devices. These types of devices have a wide range, including mobile phone, smart TVs, tractors, robots, sensors, and wearable devices.
- *Achieving ultralow latency*: The future mobile network should support real-time communications across devices. Many IoT devices such as medical equipments require very low latency in communication due to its critical importance.
- *Efficiently utilizing spectrums*: Currently, the spectrum channel is often underutilized. It is necessary to develop techniques to increase the utilization of the precious resources.
- *Catching up with bandwidth and data rate increases*: Bandwidth surge and applications with high data rate have been a major challenge to today's mobile network. This is the number one priority that 5G network should address.
- *Providing seamless connectivity across technologies*: Multiple radio technologies will coexist, and the future network should support seamless migration and roaming across technologies.

5.1.1 Architecture

To meet these requirements, the 5G employs a two-tier hierarchical structure with both the macrocell base station (MBS) and the small cell base station (SBS). Figure 5.1 shows the high-level picture of 5G. The smallest unit of coverage is constructed by various SBSs, examples of which include femtocell and picocell. The MBS oversees a set of SBSs that are located close by. For example, a building can be connected

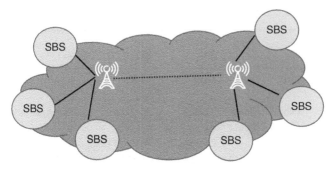

Figure 5.1 5G architecture.

by an SBS, and the SBS connects to an MBS in the region. Another example is having an SBS in the train so that all the UEs are connected to the internal SBS and will not need to deal with the frequent handovers due to fast moving of the train.

This two-tier hierarchy is the fundamental design of 5G network. It is deployed as a process of *network densification*. On the one hand, the number of antennas for each US and MBS increases. On the other hand, the density of base stations also increases. The main advantages of having SBS include the following: (1) high data rate and efficient spectrum usage; (2) reducing energy consumption; (3) SBS is a cost-effective solution; (4) the SBS is often plug and play and thus easy to manage; (5) reducing congestion on the MBS since SBS can offload the UEs; and (6) improving handoff performance because the SBS can handle handoff on behalf of its connected UEs.

The SBS can be connected to core network from the backhaul technologies. Typical backhaul technologies include wired optical fiber, wireless point to multipoint, and wireless point to point using bidirectional links. The SBS and MBS communicate with each other. The network management can be achieved by a centralized management device or using distributed method or the hybrid approach.

Table 5.1 shows the evolution of cellular networks in the past half a century. Compared to the previous generations, 5G network has much higher data bandwidth and has heavily utilized cloud-related technologies in its core network, such that it can better keep up with the demand and resource elasticity.

Since this book is related to SDN and NFV's usage in 5G network, in the following, we will focus on the core network of 5G and present how these new technologies can be used in the 5G core network. Among

Table 5.1 Cellular technology evolution.

Mobile network	1G	2G	3G	4G	5G
Deployment time	1970	1980	1990	2000	2014
Data rate	2 kbps	64 kbps	2 Mbps	200 Mbps	>1 Gbps
Technology	Analog	Digital	Broadband, CDMA, IP	IP, LAN, WAN, WLAN	4G+
Multiplexing	FDMA	TDMA	CDMA	CDMA	CDMA
Core network	PSTN	PSTN	Packet network	Internet	Cloud, SDN, NFV

the use cases, an important one is SFC, which chains multiple network functions together. We will start with the SFC discussion and then followed by the other use cases in 5G context.

5.2 Service Function Chaining

SFC enables the creation of composite (network) services that consists of an ordered set of service functions (SFs), which must be applied to packets and/or frames selected as a result of classification. SFs are network functions responsible for specific treatment of received packets. Today, these SFs are commonly performed by middle boxes. Some typical SFs are network address translation, firewall, malware detection, lawful intercept, load balancing, accounting and policy functions, traffic policing, and so on. In server environments, SFs are virtualized to operate in virtual machines (VMs) or containers. The implementation of a single SF can use a set of virtual or physical machines. Examples of SFC in telecom network are shown in Figure 5.2.

Services are constructed as sequences of SFs forming chains. These SFCs are an abstract view of a service that specifies the set of required SFs and the order in which they must be executed. A service forwarding graph is used to represent these SFCs where the nodes of the graph represent SFs. Each node of a graph can be part of one or many abstract SFCs. A single SF can appear one or more times in a given SFC. A given SFC may start at the origination point of the forwarding graph or at any subsequent node in the forwarding graph. SFs can also perform

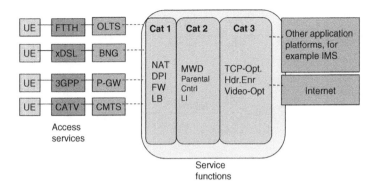

Figure 5.2 SFC in telecom network.

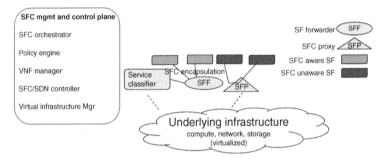

Figure 5.3 SFC framework.

branching in the service forwarding graph. It is also possible to have multiple termination points in an SFC.

There are four main components to the SFC architecture as illustrated in Figure 5.3. The main components are the service classification function (SCF), the chain termination function (CTF), the service function forwarder (SFF), and the SF. The SCF is responsible for determining the appropriate starting chain for processing each packet, encapsulating the service packets received at the classifier, and forwarding the encapsulated packets to the first SFF.

The SCF may be a very simple function (i.e., every packet received on port 1 will be processed by a particular chain) or may be very complex (i.e., use deep packet inspection to make the chain decision based on the progress of the application). We called the former static service chain and the latter dynamic service chain. Two examples of static service chain and dynamic service chain are shown in Figure 5.4.

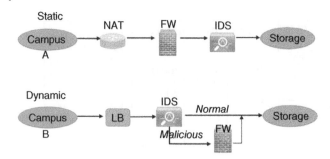

Figure 5.4 Static and dynamic SFC examples.

The SFF is the mechanism used to forward packets along the chaining path to the SF. Logically, the SFF forms a switching layer above the existing virtual networking layers. The SFs are elements that perform specific functions on the packets. SFs are different from applications since they perform intermediate or middle operations on packets rather than terminating functions. An SF may be a chain-aware SF or a chain-unaware SF. Chain-aware SFs are capable of processing and forwarding using chain-specific encapsulations and forwarding rules. The CTF is responsible for decapsulating the packet and sending it along to the destination. CTFs may perform forwarding based on the identity of the chain used to process the packet, addresses within the packet, or meta-data associated with the packet.

5.2.1 OpenFlow-Based SFC Solution

Recently, there have been some efforts on how to steer traffic to provide inline service chaining. These mechanisms are designed to explicitly insert the inline services on the path between the endpoints, or explicitly route traffic through different middleboxes according to the policies. Simple [106] proposes an SDN framework to route traffic through a flexible set of service chains while balancing the load across network functions. FlowTags [107] can support dynamic service chaining. In this section, we summarize various solutions as follows:

- *Single box running multiple services*: This approach consolidates all inline services into a single box and hence avoids the need for dealing with inline service chaining configuration of the middleboxes. The operator adds new services by adding additional service cards to its router or gateway. This approach cannot satisfy the openness requirement as it is hard to integrate existing third-party service appliances. This solution also suffers from a scalability issue as the number of services and the aggregated bandwidth is limited by the router's capacity. The number of slots in chassis is also limited.

- *Statically configured service chains*: The second approach is to configure one or more static service chains where each service is configured to send traffic to the next service in its chain. A router classifies incoming traffic and forwards it to services at the head of each chain based on the result of the classification. However, this approach does not support the definition of policies in a centralized manner and instead requires that each service to be configured to classify and steer traffic to the appropriate next service. This approach requires a large amount of service-specific configuration and is error prone. It lacks flexibility as it does not support the steering of traffic on a per subscriber basis and limits the different service chains that can be configured. Getting around these limitations would require additional configuration on each service to classify and steer traffic.

- *Policy-based routing*: A third approach is to use a router using policy-based routing and for each service to be configured to return traffic back to the router after processing it. The router classifies traffic after each service hop and forwards it to the appropriate service based on the result of the classification. However, it suffers from scalability issues as traffic is forced through the router after every service. The router must be able to handle N times the incoming traffic line rate to support a chain with $N + 1$ services.

- *Policy-aware switching layer*: Recently, it is proposed to use a policy-aware switching layer for data centers, which explicitly forwards traffic through different sequences of middleboxes. Each policy needs to be translated into a set of low-level forwarding rules on all the relevant switches. Often SDN is used for programming these policies. We will elaborate more next.

The components of StEERING [108] are shown in Figure 5.5. Our system uses a logically centralized OpenFlow-based controller to manage both switches and middleboxes. The black solid line and grey solid line in Figure 5.5 show two different service paths: the black line traversing multiple NFs and the grey line bypassing those NFs. In our design, the service paths are unidirectional, that is, different

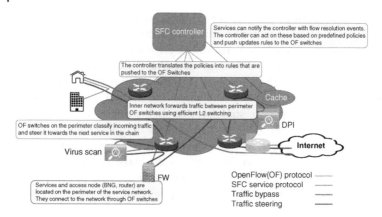

Figure 5.5 OpenFlow-based SFC solution.

service paths are specified for upstream and downstream traffic. The black solid line in this figure shows a service path for the upstream traffic through virus scan, DPI, and content cache. The grey solid line shows a service path that bypasses all the services. StEERING architecture uses two different types of switches. The Perimeter OF Switches are placed on the perimeter of the service delivery network. These switches will classify the incoming traffic and steer it toward the next service in the chain. These are the switches to which services or gateway nodes are connected. The inner switches will forward the traffic using efficient L2 switching. These switches are only connected to other switches. These switches may or may not be OF switches.

Traffic steering is a two-step process. The first step classifies incoming packets and assigns them a service path based on predefined subscriber and application. The second step forwards packets to a next service based on its current position along its assigned service path. This two-step traffic steering process only needs to be performed once between any two border routers, regardless of the number of switches that connects them.

We define two types of traffic steering functionalities as follows, depending on the information needed for identifying the steering policies.

Basic traffic steering is based only on the Layer 3 and Layer 4 (L3–L4) headers, without any deeper inspection of packets. The policy is derived from subscriber and application policies. It does not involve any per-flow forwarding rules or state. Rather, it involves setting up forwarding rules a priori, which are derived from configuration and can apply to an aggregate of many flows (e.g., all flows to a particular IP address block and/or port number). Basic traffic steering includes per-subscriber and application configuration, so that different services may be applied according to individual subscriber preferences, as well as the other L3–L4 criteria described earlier. Service chains defined by the basic traffic steering mechanism are in effect starting from the first packet of a flow (e.g., TCP SYN packet). Therefore, services that are bypassed using basic steering do not see this first packet.

Complex traffic steering is based on installing per-flow forwarding rules dynamically, in response to the first few packets of the flow being processed by the service. This implies the possibility of using deep packet inspection to analyze L5–L7 flow contents. Allowing more granular control over traffic flows based on higher-level application information, such as URLs, is a key advantage of complex steering. These per-flow rules are installed by the OF controller in response to notifications from an inline service such as DPI. Since these rules are installed reactively rather than a priori, and especially since TCP flow contents are not available for inspection by DPI until after the three-way handshake, the first packets of a flow will traverse a different initial service chain before a final service chain is established for that flow. So, any services that are only bypassed via complex steering will see the first packet.

The inline services may (optionally) notify the OF controller that certain content or metrics have been identified via the StEERING Service Protocol, shown as the grey dotted lines in Figure 5.5. For instance, once DPI has recognized or resolved a flow, it can send a notification to the OF controller.

The data plane (forwarding) can be easily configured and scale as the number of subscriber/application combination grows. The controller programs switches with the rules on how to forward each packet. Forwarding decisions are made based on Layer 2 to Layer 4 contents of packets as well as the ingress port. The key challenge to achieve scalability is to avoid exponential growth (rule explosion) of the forwarding rules installed in each switch. We make three design choices to reduce the amount of state on each switch: defining port types to indicate directions, using multiple tables to decompose multidimensional policies, and introducing a new metadata type to encode service paths.

5.2.1.1 Represent Directions with Port Types

We define two types of ports on perimeter switches: *node* ports and *transit* ports. Node ports are connected to services and gateway nodes (BNG, GGSN, routers). Transit ports are connected to other perimeter switches or to inner network switches.

Figure 5.6 shows an example of a service delivery network based on the StEERING architecture. Switches OF1, OF2, and OF3 are perimeter switches. Switch SW1 is an inner switch. The black/grey ports on the switches are node ports and the white ports are transit ports.

Incoming traffic, either coming in from a gateway node or coming back from a service, always enters the service delivery network via a perimeter switch and through a node port. Packets coming in through node ports are steered toward the next node (service or gateway) in their assigned service paths. Packets arriving on transit ports are simply forwarded using their destination MAC address.

All packets traversing the steering network are considered to be traveling either upstream or downstream. Each node port in the steering network is either facing upstream or downstream. In Figure 5.6, downstream-facing ports are colored grey and upstream-facing ports are colored black. All packets that arrive on a downstream-facing port are traveling upstream, and vice versa. Packets arriving on transit ports may be traveling in either direction. In this case, the direction is known based on the destination MAC address, which will correspond to either an upstream-facing or downstream-facing service or router port.

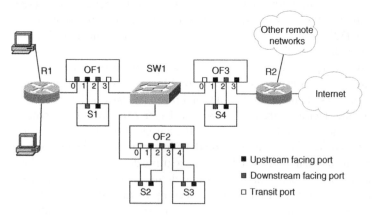

Figure 5.6 Illustration of port direction.

5.2.1.2 Realize Policies with Multiple Forwarding Tables

In theory, a single TCAM-like table could be used to specify the required functionality, as in OpenFlow 1.0 or pswitch. However, this would not be a scalable solution because it would involve the cross-product of subscribers, applications, and ports. *Using indirection and multiple tables, we separate this into multiple steps, resulting in linear scaling of each table.*

Six mandatory tables are used in each OF switch, as shown in rectangular in Figure 5.7. We introduce each table as follows.

The first table is the direction table. It uses the ingress port as the key, and serves two purposes: to determine whether the packet arrived on a node port or a transit port, and in which direction the packet is headed.

The key for the MAC table is the packet's destination MAC address. Based on the contents of this table, the packet will either be transmitted to a directly connected service or router on a node port, forwarded out to another transit port, or dropped.

The next table is the subscriber table. It is used to get a subscriber's default service set for the current direction. The key is the direction bit together with the subscriber's IP address. The subscriber's IP address comes from either the source or destination IP address fields, depending on the direction. This can be a longest-prefix match (LPM) table. If there is a miss in this table, the default action is to drop the packet.

Following the subscriber table is the application table. In this context, "application" refers to the remote communication endpoint, for example, web servers, as identified by the IP address, protocol, and port number. It is used to modify the subscriber's default service set according to any static L3–L4 application policies. Wildcards, prefixes, and ranges are permitted in this table. Based on this information, specific services can be excluded from the service set or added to it.

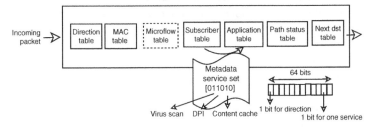

Figure 5.7 Multiple tables and metadata.

If there is a miss in this table, the packet is not dropped and the service set is not modified.

The path status table follows the application table or Microflow table (will be described in the following section). Its purpose is to determine which services in the service set have already been applied. This is important because a packet may traverse the same perimeter switch multiple times, and it should be treated differently each time. The ingress port is sufficient to provide this information. If this table is reached, it means that the packet has arrived on a node port, connected directly to a service or router. The ingress port then tells us which service was just applied, if any, and it also tells us the direction. There is a global ordering of services in each direction (they may or may not be the exact reverse of each other). Based on the direction and the previous service, the service set is modified to exclude the previous service and all other services that precede it.

The final table along the node port path is the next destination table. It uses the direction and the service set as a key. This is a TCAM-like table, with arbitrary bitmasks and rule priorities. Based on the direction bit, it essentially scans the bits in the service set according to the global service ordering in that direction. The first or highest-priority service it finds will be the next destination. If the service set is empty, the next destination will be either the upstream or downstream router, depending on the direction bit. The next destination may be connected to the current switch or another one. If the destination is connected to a different switch, then the destination MAC address is set to the value corresponding to that service or router and the packet is transmitted out through an appropriate transit port. If the destination is directly connected, then the MAC addresses are updated as needed and the packet is transmitted out to the corresponding node port.

5.2.1.3 Handle Dynamics with the Microflow Table

The Microflow table is added to handle dynamically generated rules. We assume that operators have static policies for the subscribers and applications. Therefore, the subscriber table and the application table can be programmed with static rules. But in real time, operators may want to add policies dynamically, or add more specific policies, or higher priority policies. Moreover, policies may be added according to the results of another middlebox, for example, DPI. We add the Microflow table to accommodate such dynamics.

If there is a hit in the direction table, the next table to be consulted will be the Microflow table. The key for this table is the direction bit together with the 5-tuples (source and destination IP address, IP protocol field, and TCP/UDP source and destination port) of the packet. The table contains exact-match entries used for selective complex steering of specific TCP/UDP flows. If there is a hit in this table, the next two lookups will be skipped. Thus, the rules in the Microflow table have higher priority than the rules in the subscriber and application tables.

5.2.1.4 Encode Service Chaining with Metadata

Metadata is used in OpenFlow 1.1 to communicate the information among different tables and associated actions [109]. There are two types of metadata required in data paths. Some part of metadata is used as lookup action in further tables and some metadata is required by actions. Intermediate results from one table are communicated to other tables using some metadata, which can be used as part of a subsequent lookup key or be further modified later.

We introduce two new types of metadata, the direction bit that represents the direction of the flow and the set of inline services to be applied for the flow under process, called *service set*. This service set is encoded as a bit vector, one bit for each possible service. More sophisticated encodings can be used to enable more advanced features such as load balancing over multiple service instances. OpenFlow 1.1 supports 64-bit metadata field [109]. This requires one bit for the direction and leaves up to 63 bits for encoding the service set, allowing a maximum of 63 distinct services. The format of the metadata is shown in Figure 5.7, together with an example. The metadata field can be applied with arbitrary mask and is updated as follows:

$$new\_meta = (old\_meta \& \neg mask) \| (value \& mask) \qquad (5.1)$$

The *value* are the bits to be set and the mask is used to select these bits, which can be arbitrary 64 bit numbers.

The first bit indicates whether it is upstream (0) or downstream (1). The following N-bit vector defines the service set, encoding for N number of services. The encoding in this example specifies that this packet should traverse the virus scan, DPI, and content cache. In the data path, this metadata is set, then modified, and finally used to search for the next service to be applied to this packet.

5.2.1.5 Summary of Dataplane Functions

Figure 5.8 depicts the details of the forwarding steps in the data path. When a packet arrives, using its input port, a lookup in the direction table determines if this packet needs to be classified at this hop or should be just transmitted to the next hop. If it is the latter, the packet is sent out according to its MAC address by performing a lookup in the MAC table. If a classification is needed at this hop to determine its service sets, then lookups are performed at the Microflow, subscriber, and application tables. Each table can independently operate on the service set. The lookup keys are specified in the brackets below the table names in the figure. According to the bits set in the service set, the path status table determines which services have already been traversed, and what is the next service in the chain. Finally, the packet is set with the right MAC address in the next destination table and sent out to the corresponding output port.

5.2.2 SFC Monitoring

Regardless of what mechanism is used to implement the service chaining, one important problem is how to verify that the path has been correctly installed. The goal is to prove that packets of a given flow have traversed the expected path. Existing reachability measurement includes ping and traceroute to measure the reachability from a source to a destination. Ping triggers ICMP replies and traceroute triggers ICMP TTL expiration messages on the routers along the path. Both methods do not require two-end control. There has been ping and traceroute at different protocol layers, for example, MPLS ping.

However, as stated earlier, the traditional ping/traceroute method is not suitable for the inline service setting. In traditional network, the loss of ping/traceroute packets indicates the path problem. However, in our setting, the ping/traceroute packet may not be recognized by the service (middlebox) in the middle of the path, and thus got dropped. Thus, we cannot simply say the symptom of lost measurement packets is due to the path reachability problem. Therefore, we need a different method to measure the path reachability for inline service chaining.

We define the reachability problem of inline service chaining as follows. Assuming that a flow f traverses service chain (S1, S2, and S3) in order, the network topology is shown in Figure 5.9. In this example, the services are connected to switches F1, F2, and F3, respectively. The flow enters from the ingress switch F0 and exits the network from

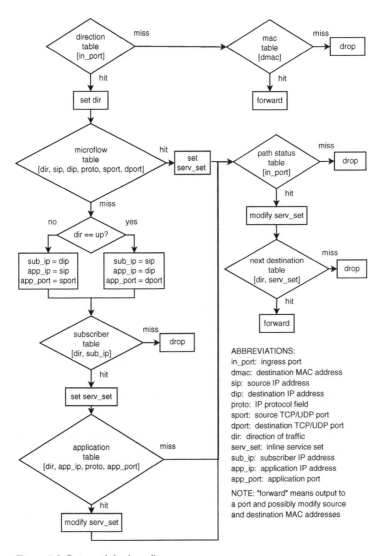

Figure 5.8 Data path lookup diagram.

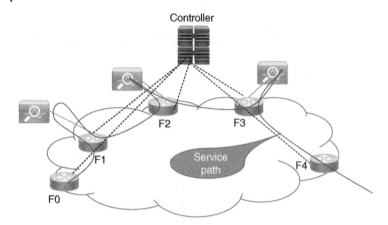

Figure 5.9 Example to show service chain reachability.

egress switch F4. The goal is to test that flow f has traversed the path
$(F0 - F1 - S1 - F1 - F2 - S2 - F2 - F3 - S3 - F3 - F4)$. The reachability implies three dimensions as follows:

1) The service path is set up correctly; the traffic traverses service S1, S2, and S3 in this order.
2) There is no failure at the forwarding path, that is, switches F0–F4 function correctly to forward the packets.
3) There is no failure in the services, that is, services S1–S3 are alive and send/receive packets correctly.

What it does not imply is that services S1–S3 perform the right action to the packet. That is, we do not test if the service is functioning correctly, since this highly depends on the service logic and the knowledge of the service.

From the description above, we can see that some services may discard packets as a part of their functions, for example, firewalls to stop flows, rate limiters to drop packets, which may require more careful interpretation to the monitoring results. We propose that the monitoring system analysis module should be aware of the functional logic of each service and take this into consideration when interpreting the results.

Similarly, it is possible that some services modify the packet header fields, such as NAT. If the controller can accurately know the modification, it can create a mapping of the flows before and after a service and still capture the packets for the same flow. If not, then the controller can use a combination of packet headers and payloads as the key to capture packets.

5.2.2.1 Handling Multiple Monitoring Tasks

In the second step, we tested that the chain has been implemented correctly. In the third step, we need to continuously monitor the chain to make sure there is no failure in any component of the network.

Our key idea is that in this step, we do not need to monitor after every service in the chain in the second step. Instead, we can monitor a small number of points in the service chain, or even just monitor the two ends of a service chain. For example, as shown in Figure 5.10, there are two service chains, the grey one and the black one.

We define a monitor point to be the point where we install mirroring rules. Figure 5.10 shows a set of example monitor points in grey circles. With monitor points of A and B, we can ensure if there is any failure in any of the services or switches in this part of chain segment. Thus, given this set of monitor points A–E, if any failure occurs in service S1, we can detect a reachability failure from monitor point B. If any failure occurs in service S2 and S3, we will detect it from both monitor C and monitor E. If we observe that both C and E are affected, but B is not affected, then it suggests likely that either S2 or S3, or any of forwarding elements along this path segment fails. This example shows that

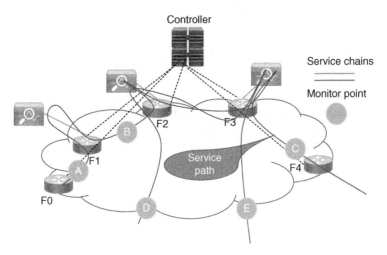

Figure 5.10 Example to show monitor points for multiple service chains.

Algorithm 1 Monitor point selection algorithm

---

procedure *Select_Monitor_Points(P,G)*

    create a bipartite graph N with two sets of nodes for all elements in P and G

    for each flow $p \in P$ **do**

      for each element $f \in p$ **do**

        add edge $f \rightarrow p$

      end for

    end for

    Sort element $f \in N$ according to its out degree

    for each element $f \in N$ **do**

      select f to be a monitor point, remove p

      if $p \in N = \emptyset$ **then**

        break;

      end if

    end for

---

the paths for multiple service chains may overlap, and we can select a subset of the points in the path to monitor, in order to reduce the load for continuous monitoring.

The problem is defined as follows. A service chain is defined as a sequence of forwarding switches and the services, for example,

p0=F0,F1,S1,F2,S2,F3,S3,F4,
p1=F1,F2,S2,F3,S3,F5.

Given multiple service chains, P=p0,p1, …, pk, we need to select a minimum number of monitor points so that all the path segments are covered.

Our key idea is to create a bipartite graph by mapping each monitor point candidates to the service chain it can cover. Then, the monitor points are sorted according to the out degree (the number of service chains it can monitor). We finally greedily select the monitor points until all the services are covered. The detailed algorithm is shown in Algorithm 1.

5.2.3 Optical SFC

Optical communications have already enabled terabits-level high-speed transmission. One promising technology is the dense wavelength-division multiplexing (DWDM). It allows a single fiber to carry tens of wavelength channels simultaneously, offering huge transmission capacity and spectrum efficiency. On the other hand, reconfigurable wavelength switching devices have already been widely deployed in long-haul and metro transport networks, providing reconfiguration on layer-0 lightpath topologies. We argue that optics can be used in today's DCs to the end-of-rack switches, the top-of-rack switches, as well as the servers. Although switching in the optical domain may have less agility than the packet-based approaches, it is suitable for the dynamic level required by service chains consisting of high-capacity core NFs and use of traffic aggregation.

We propose that optical technology can be used to support traffic steering. In the following, we present a packet/optical hybrid DC architecture, which enables steering large aggregated flows in an optical steering domain. Figure 5.11 illustrates an overview of the proposed architecture. The centralized OSS/BSS module interfaces with an SDN controller and a cloud/NFV manager.

The cloud manager is responsible for cloud resource allocation and automating the provisioning of VMs for VNFs. It also includes an NFV management module that handles instantiation of the required VNFs while ensuring correctness of configuration.

The SDN controller can be a part of the cloud management subsystem or a separate entity. The SDN controller and cloud/NFV manager perform resource provisioning. On the southbound interface, the SDN controller uses optical circuit switching to control the network elements. This interface can be realized using OpenFlow v. 1.4, which has an extension for optical circuit configuration.

To perform service chaining, the OSS/BSS module needs to request VNFs and network resources and the policies for resource allocation. The SDN controller performs network-resource allocation by relying on a path computation entity that could be integrated with the SDN controller.

The data center contains both an optical steering domain and a packet steering domain. The optical steering domain conducts traffic steering for large aggregated flows. The entry point is an ROADM, which either forwards a wavelength flow to the optical domain or sends it to the packet domain for fine-grained processing. After a wavelength flow has gone through the needed VNFs, it is steered back to the ROADM. The flow is controlled to route either back to the packet domain for fine-grained processing, or forward to the optical domain for high-capacity optical processing.

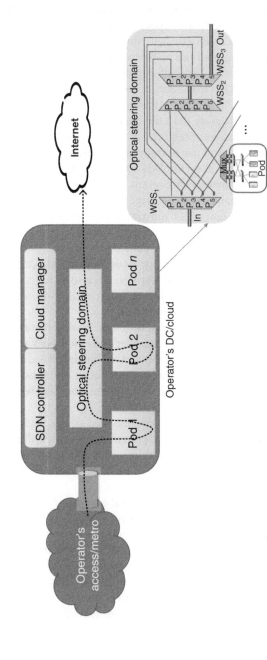

Figure 5.11 Optical service function chaining framework.

5.2.3.1 Service Placement in Optical NFV

As shown above, optical switching and packet switching can be jointly employed to simultaneously address the efficiency and the flexibility requirements of VNF chaining. Optical switching serves as the "backbone" of a DC network to steer large flows to pods in the form of wavelength. Upon arrival to the pod, the flow is converted to packet traffic, and steered across the servers where the needed VNFs are hosted. The flow will need to be converted back to wavelength after it has been processed by all the needed VNFs in that given pod. For each pod the traffic of a given VNF chain visits, an O/E/O conversion is needed between the pod and the optical steering domain. Note that for small flows, it could be more efficient to go through their service chain in a pure packet form, for example, a parallel packet steering domain that handles small flows.

The data plane of the referenced architecture consists of an operator's access/metro networks, and an operator's DC/cloud. Traffic flows destined to the same set of core VNFs are aggregated before they are routed to the DC/cloud. This process is based on the operator's policy and prior packet classification using deep packet inspection. Conventional packet technologies can be employed, for example, multiprotocol label switching (MPLS) or an OpenFlow-based scheme with packet rules defined. The aggregated flow is then converted by tunable transceivers into a wavelength flow following the wavelength division multiplexing (WDM) standard. Multiple wavelength flows can be multiplexed into the same fiber for transmission efficiency.

O/E/O conversion is an expensive and power-intensive process that often calls for minimization in traffic engineering and network resource planning problems. In the referenced architecture, given sufficient resources, it is desirable to instantiate all the VNFs of an NF chain within a single pod. This strategy will avoid unnecessary flow traversals in the optical steering domain. In addition, resource allocation within a single pod can help reduce the complexity of resource orchestration in both the DC network and pods' resources. However, in reality, it may not always be possible to provision an entire NF chain within a single pod due to resource constraints. In a trace published by Google [5], the required resources may include computation, storage, networking, and so on. When an NF chain has to visit multiple pods, VNF placement will affect the number of required O/E/O conversions.

Figure 5.12 shows an example of two different VNF placements for three NF chains. Placement 1 is better because all the NFs of NF

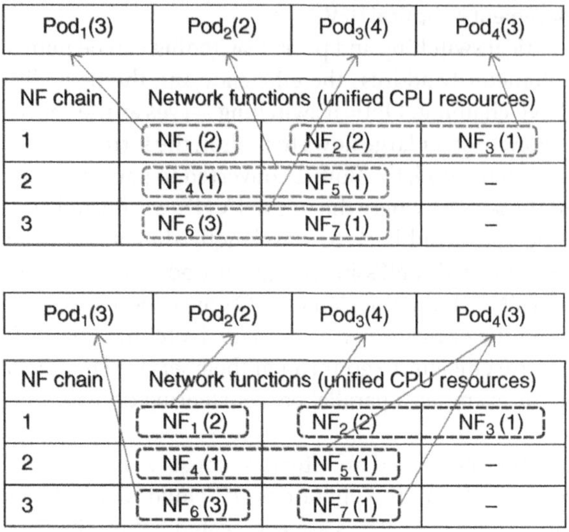

Pod$_1$(3)	Pod$_2$(2)	Pod$_3$(4)	Pod$_4$(3)
NF chain	Network functions (unified CPU resources)		
1	NF$_1$(2)	NF$_2$(2)	NF$_3$(1)
2	NF$_4$(1)	NF$_5$(1)	–
3	NF$_6$(3)	NF$_7$(1)	–

Pod$_1$(3)	Pod$_2$(2)	Pod$_3$(4)	Pod$_4$(3)
NF chain	Network functions (unified CPU resources)		
1	NF$_1$(2)	NF$_2$(2)	NF$_3$(1)
2	NF$_4$(1)	NF$_5$(1)	–
3	NF$_6$(3)	NF$_7$(1)	–

Figure 5.12 Optical service function chaining placement problem. (a) NF placement 1 (better) and (b) NF placement 2.

Chain 3 are placed in Pod3, while NF Chain 3 in Placement 2 needs to visit two pods, resulting in one more O/E/O conversion. This simple example shows how VNF placement can minimize O/E/O conversions. The problem of VNF placement can be stated as follows.

Each NF chain has a set of VNFs to be placed in a number of pods. The VNFs of the same chain placed in the same pod form a placement group. Eventually, the VNF set is partitioned into one or multiple placement groups. The optimization objective is to minimize the total number of placement groups for all the NF chains, respecting the resource constraints.

A placement group of an NF chain corresponds to a pod the NF chain needs to visit, and thus one O/E/O conversion is needed. We make the assumption that a single VNF will never be split into more than one pod. We also assume sufficient pods, such that all the NF chains can be accommodated. A variant of this problem can be minimizing NF blocking given a constraint on the number of pods, which will be investigated in our future study. One drawback of minimizing O/E/O conversions is that it may limit the freedom of the cloud manager to choose a specific pod for an NF chain (e.g., for affinity requirement). Such cases can be addressed by adding specific extra

---

Algorithm 2 Optical NFV Placement algorithm

---

procedure *Place(PodList, ChainList, NFList*

  create a bipartite graph N with two sets of nodes for all elements in P and G

  for each NF chain c in ChainList **do**

    Initiate an empty pod list PodListc.

    Sort the NFs needed by c in a descending order by resource demand

  end for

  for each unprocessed NF n of c **do**

    Set flag=false

    Sort PodListc in an ascending order by available resource and select the first pod p

    if p has enough resources for n **then**

      provision n by p and set flag=true

    else

      go to the next pod p in PodListc

    end if

    if flag=false **then**

      add the most-available-resource pod pm of PodList to PodListc. Provision n by pm.

    end if

  end for

  In PodList, select pod pa with the least resources consumed by NFs provisioned in Step 3, and pod pb with the most available resources. Move all the NFs of pa to pb if resources allow and continue on 4a, otherwise end the algorithm.

---

constraint or by simply having the operator manually assign VNFs for the chain in question.

According to the problem statement above, we formulate the problem of NF placement as integer linear programming (ILP). Since the ILP formulation has only binary variables, it becomes binary integer programming (BIP). In this formulation, we assume integrity of NF, that is, an NF cannot be split into more than one pod.

However, the BIP-based solution does not scale with the size of the input (e.g., C, M, and N); therefore, we design a heuristic algorithm for computation efficiency, shown in Algorithm 2. Similar to BIP, Algorithm 2 takes CPU as the most limiting resource and may conduct additional checks to ensure no violation of other resource constraints. Due to the space limitation, we do not include this part in Algorithm 2. The high-level idea of Algorithm 1 is, VNF placement for chain c follows a "best-fit" strategy, that is, for each NF chain c, the required NFs are first sorted by their resource demand in a descending order. Next, each time the algorithm selects an unplaced NF of chain c with the highest resource demand, and places it to the pod in PodListc with the least (but sufficient) CPU resources. PodListc is initialized as an empty list for NF chain c, and records all the pods used by c. If PodListc is empty, or no such pod can be found in PodListc, the algorithm selects the pod with the most resources from the entire pod list PodList, add it to PodListc, and repeat VNF placement in Step 3. In this study, we assume sufficient pods to avoid NF blocking. A variant of the problem can be minimizing the block ratio of NF placement, given a limited number of pods. In Step 3, as pod usage is optimized per NF chain basis, this does not automatically lead to minimization of overall pod number. Therefore, we introduce Step 4, which conducts an additional optimization to consolidate pods for all NF chains.

5.2.4 Verification of Service Function Chaining

The traditional approach to detecting configuration errors is to apply network verification. Unfortunately, current network verification techniques focus on verifying the flow table rules in Layer 2 and Layer 3 devices against basic invariants such as loop freeness and no blackholing [74, 110]. These approaches assume a simple stateless forwarding model where the action solely depends on a match on the header fields of each packet. However, many NFs are stateful; the handling of a packet depends not only on the current packet but also on previously encountered packets, in some cases, on the content of packets. For example, a firewall, discussed earlier, allows packets from external hosts only if it has previously seen an outgoing packet from internal host for that connection.

One way to extend verification to NFs is using test packets to probe the NF and determine the set of invariants that are satisfied or violated given the NF's current state. These active probing techniques can detect problems after they are deployed, but we still need static verification to capture the problems before the deployment. In addition, these techniques require access to the source code but, due to proprietary reasons, most NFs are closed source. An alternate approach is not to rely on the source code. Further, to enable Telcos to continue to provide carrier-grade high availability and reliability, it will be crucial to verify the correctness of NFs and service chains before deployment and not after.

Performing verification for NFs and their service chains introduces several interesting challenges. First, NFs maintain state about each flow and perform different actions based on the states. The states also vary largely depending on the type of NF being verified. Existing verification tools fail to support it because they are built on the OpenFlow type of forwarding abstraction, where all packets of a flow are handled the same using a match-action rule. Instead, stateful SFC verification requires a new forwarding abstraction to consider the disparate state for individual flows (or connections).

Second, while verifying the functionality of one NF is tractable, networks consist of many NFs that are chained together. To verify a service chain, a framework must verify not just one NF but all NFs and switches that the flows will traverse: essentially, verifying the entire network. While verifying stateless network devices is already challenging [74, 110], adding stateful devices further complicates the problem. Thus, care must be taken to ensure that verification includes accurate and scalable algorithms that can be applied to realistic networks.

It is the simple switch forwarding abstraction, for example, Open-Flow, that makes the data plane verification feasible [110] as it hides the complexity in control plane and represents a unified interface to the verification. In the context of NFV, unfortunately there is no such simple abstraction to capture the data plane behavior of an NF. Inspired by the OpenFlow match-action abstraction, we can use a new NF forwarding abstraction. It comprises of two parts: a match-action table and a state machine. The state machine is a natural representation of stateful processes. The nodes in the state machine are the states each NF maintains, and the edges capture the conditions that trigger state transitions. The table contains match-action rules: matching on both the packet header and the internal states, performing the action on packets, and changing the internal states. Illustration of the stateful table is shown in Figure 5.13. Use expert knowledge or by talking with network operators, we can determine the state machines for the NFs.

We observe that all existing verification tools build some form of forwarding graphs from the rules and then perform reachability checks on this graph [110], as graph traversal algorithms are efficient and scalable. However, this approach is insufficient for our problem because

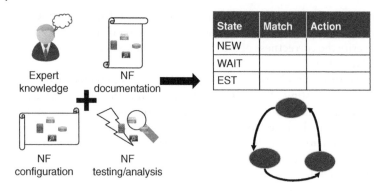

Figure 5.13 Model used for SFC verification.

the forwarding graphs only capture the forwarding behavior of the network but not the state transitions of the NFs. On the one hand, receiving and forwarding a packet may trigger the NF's state transition. On the other hand, the state changes may affect the forwarding behavior of subsequent packets of the same flow. Thus, naturally we should combine them in order to correctly verify the stateful network behavior. We can use a stateful forwarding graph (SFG) that encodes both the state transitions and forwarding behavior. We develop an algorithm that automatically generates the SFG from our NF tables and FSMs. Figure 5.14 shows an example for the SFG. In an SFG, each node is denoted as $\langle H, D, S \rangle$, representing any packet in the packet header space H arriving at a network device (switch or NF) D, when the network device is in a particular state S.

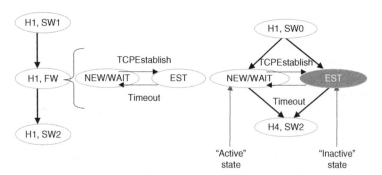

Figure 5.14 Stateful forwarding graph for SFC verification.

An edge pointing from one node $\langle H_1, D_1, S_1 \rangle$ to another $\langle H_2, D_2, S_2 \rangle$ means when a packet in H_1 arrives at D_1 with state S_1, it will be modified to H_2 and forwarded to a device D_2 at state S_2. If the device D_1 does not modify the packet header, then H_1 is equal to H_2. If the packet H_1 does not trigger the state transition, then S_1 is equal to S_2. Figure 5.14 shows an example of a path of packets H_1 traversing through D_0, D_1, and D_2.

Essentially, an SFG edge can represent one of two events: (1) a packet is modified and forwarded to next hop or dropped; (2) a device is changing its state to the next state. Accordingly, we use two different types of edges to represent the different events.

Using the SFG, we develop a verification algorithm that is capable of verifying the dynamic and stateful behavior of the network. The verification leverages the structure of our NF abstract model, namely due to the use of state machine, for each flow the NF will be in *one and only one state at any given time*. Essentially, only one node in an NF's state will be active. By activating different nodes (corresponding to different states of an NF) during the verification process, we are able to verify different forwarding scenarios across NFs and states.

5.3 Core Network Functions Virtualization: vEPC

The 3G packet core (PC) network consist of three interacting domains: core network (CN), 3G PC terrestrial radio access network (UTRAN), and UE. The main function of the core network is to provide switching, routing, and transit for user traffic. Core network also contains the databases and network management functions. It is the common packet core network for GSM/GPRS, WCDMA/HSPA, and non-3GPP mobile networks. The packet core system is used for transmitting IP packets.

The core network is divided into circuit- and packet-switched domains. Some of the circuit-switched elements are mobile switching center (MSC), visitor location register (VLR), and gateway MSC. Packet-switched elements are SGSN and GGSN. Some network elements such as EIR, HLR, VLR, and AUC are shared by both domains.

The architecture of the core network may change when new services and features are introduced. Number portability database (NPDB) will be used to enable user to change the network while keeping their old phone number. Gateway location register (GLR) may be used to optimize the subscriber handling between network boundaries. The primary functions of the packet core with respect to mobile wireless networking are mobility management and QoS. These functions are not typically provided in a fixed broadband network, but they are crucial for wireless networks. Mobility management is necessary to ensure packet network connectivity when a wireless terminal moves from one base station to another. QoS is necessary because, unlike fixed networks, the wireless link is severely constrained in how much bandwidth it can provide to the terminal, so the bandwidth needs to be managed more tightly than in fixed networks in order to provide the user with an acceptable quality of service.

The signaling for implementing the mobility management and QoS functions is provided by the GPRS tunneling protocol (GTP). GTP has two components:

- *GTP-C*: A control plane protocol that supports establishment of tunnels for mobility management and bearers for QoS management that matches wired backhaul and packet core QoS to radio link QoS.
- *GTP-U*: A data plane protocol used for implementing tunnels between network elements that act as routers. There are two versions of GTP-C protocol, that is, GTP version 1 (GTPv1-C and GTPv1-U) and GTP version 2-C (designed for LTE). In this invention, we focus on GTPv1 and the 3G PC-based system.

Network services are considered end to end; this means from a terminal equipment to another. An end-to-end service may have a certain QoS that is provided for the user of a network service. It is the user that decides whether he/she is satisfied with the provided QoS or not. To realize a certain network, QoS service with clearly defined characteristics and functionality is to be set up from the source to the destination of a service. In addition to the QoS parameters, each bearer has an associated GTP tunnel. A GTP tunnel consists of the IP address of the tunnel endpoint nodes (radio base station, SGSN, and GGSN), a source and destination UDP port, and a tunnel endpoint identifier (TEID). GTP tunnels are unidirectional, so each bearer is associated with two TEIDs, one for the uplink and one for the downlink tunnel. One set of GTP tunnels (uplink and downlink) extends between the radio base station and the SGSN, and one set extends between the SGSN and the GGSN. The UDP destination port number for GTP-U is 2152, while the destination port number for GTP-C is 2123. The source port number is dynamically allocated by the sending node.

5.3.1 Existing Solutions Problems

The 3GPP standards do not specify how the packet core should be implemented; they only specify the network entities (SGSN, etc.), the functions each network entity should provide, and the interfaces and protocols by which the network entities communicate. Most implementations of the packet core use servers or pools of servers dedicated to a specific network entity. For example, a pool of servers may be set up to host SGSNs. When additional signaling demand requires extra capacity, an additional SGSN instance is started on the server pool, but when demand is low for the SGSN and high for, for example, the HSS, the HSS servers will be busy while the SGSN servers may remain underutilized. In addition, server pools that are underutilized will still consume power and require cooling even though they are essentially not doing any useful work. This results in an additional expense to the operator.

An increasing trend in mobile networks is for managed services companies to build and run mobile operator networks, while the mobile operator itself handles marketing, billing, and customer relations. Mobile operator-managed service companies may have contracts with multiple competing operators in a single geographic region. A mobile operator has a reasonable expectation that the signaling and data traffic for their network is kept private and that isolation is maintained between the traffic for their network and for that of their competitors, even though their network and their competitors' networks may be managed by the same managed service companies. The implementation technology described earlier requires the managed services company to maintain a completely separate server pool and physical signaling network for each mobile operator under contract. The result is that there is a large duplication of underutilized server capacity, in addition to additional power and cooling requirements, between the operators.

The packet core architecture also contains little flexibility for specialized treatment of user flows. Although the architecture does provide support for QoS, other sorts of treatment involving middleboxes, for example, specialized deep packet inspection or data caching for transcoding or augmented reality applications, is difficult to support with the current PC architecture. Almost all such applications require the packet flows to exit through the GGSN, thereby being detunneled from GTP, and be processed within the wired network.

5.3.2   Virtualization and Cloud-Assisted PC

The basic concept of bringing virtualization and cloud to PC is to split the control plane and the data plane for the PC network entities and to implement the control plane by deploying the EPC control plane entities in a cloud computing facility, while the data plane is implemented by a distributed collection of OpenFlow switches. The OpenFlow protocol is used to connect the two, with enhancements to support GTP routing, while the PC already has a split between the data and control plane, in the sense that the HLR, HSS, AuC are pure control plane. The EPC architecture assumes a standard routed IP network for transport on top of which the mobile network entities and protocols are implemented.

The split proposed in this document is instead at the level of IP routing and MAC switching. Instead of using L2 routing and L3 internal gateway protocols to distribute IP routing and managing Ethernet and IP routing as a collection of distributed control entities, this document proposes centralizing L2 and L3 routing management in a cloud facility and controlling the routing from the cloud using OpenFlow. The standard 3G PC control plane entities, SGSN, GGSN, HSS, HLR, AuC, VLR, EIR, SMS-IWMSC, SMS-GMSC, and SLF are deployed in the cloud. The data plane consists of standard OpenFlow switches with enhancements as needed for routing GTP packets, rather than IP routers and Ethernet switches. At a minimum, the data plane traversing through the SGSN and GGSN and the packet routing part of the NodeB in the E-UTRAN require OpenFlow enhancements for GTP routing. Additional enhancements for GTP routing may be needed on other switches within the 3G PC depending on how much fine-grained control over the routing an operator requires.

The packet core control plane parts of the gateways for GTP-C communications, that is, the parts that handle GTP signaling, are implemented in the cloud as part of the OpenFlow controller. The control plane entities and the OpenFlow controller are packaged as VMs. The API between the OpenFlow control and the control plane entities is a remote procedure call (R3G PC) interface. This implementation technology is favorable for scalable management of the control plane entities within the cloud, since it allows execution of the control plane entities and the controller to be managed separately according to demand. The cloud manager monitors the CPU utilization of the 3G PC control plane entities and the control plane traffic between the PC control plane entities within the cloud. It also monitors the control plane traffic between the UEs and NodeBs, which do not have control plane entities in the cloud, and the PC control plane entities. If the 3G PC control plane entities begin to exhibit signs of overloading, such as utilizing too much CPU time, or queuing up too much traffic, the overloaded control plane entity requests that the cloud manager start up a new VM to handle the load. The cloud manager also provides reliability and failover by restarting a VM for a particular control plane function if any of the PC control plane entities should crash, collecting diagnostic data, saving any core files of the failed PC control plane entity, and informing the system administrators that a failure occurred. The control plane entities maintain the same protocol interface between themselves as in the standard 3GPP 3G PC architecture.

The OpenFlow control plane, shown here as a gray dotted line, manages the routing and switching configuration in the network. The OpenFlow control plane connects the SGSNs, the standard OpenFlow switches, and the GGSN to the OpenFlow controller in the cloud. The physical implementation of the OpenFlow control plane may be as a completely separate physical network, or it may be a virtual network running on the same physical network as the data plane, implemented with a prioritized VLAN or with an MPLS label-switched path or even with a GRE or other IP tunnel. The OpenFlow control plane can, in principle, use the same physical control plane paths as the GTP-C and other mobile network signaling. The SGSN-Ds and the GGSN-Ds act as OpenFlow GTP-extended gateways, encapsulating and decapsulating packets using the OpenFlow GTP extensions.

The NodeBs have no control plane entities in the cloud because the RAN signaling required between the RNC and the NodeB includes radio parameters, and not just IP routing parameters. Therefore, there is no OpenFlow control plane connection between the OpenFlow controller in the cloud and the NodeBs. The NodeBs can, however, act as OpenFlow GTP-extended gateways by implementing a local control to data plane connection using OpenFlow. This allows the packet switching side of the NodeBs to utilize the same OpenFlow GTP switching extensions as the packet gateways.

The operation of the PC cloud works as follows. The UE, NodeB, SGSN, and GGSN signal to the HLR, HSS, AuC, SMS-GMSC using the standard EPC protocols, to establish, modify, and delete GTP tunnels. This signaling triggers procedure calls with the OpenFlow controller to modify the routing in the EPC as requested. The Open-Flow controller configures the standard OpenFlow switches, the Openflow SGSN, and GGSN module with flow rules and actions to enable the routing requested by the control plane entities. Details of this configuration are described in the following section.

Figure 5.15 illustrates how PC peering and differential routing for specialized service treatment are implemented. These flow rules steer GTP flows to particular locations. The operator, in this case, peers its PC with two other fixed operators. Routing through each peering point is handled by the respective GGSN1-D and GGSN2-D. The bottom two lines with arrow shows traffic from a UE that needs to be routed to either one or another peering operator. The flow rules and actions to distinguish which peering point the traffic should traverse are installed in the OpenFlow switches and gateways by the OpenFlow controller. The OpenFlow controller calculates these flow rules and actions based on the routing tables it maintains for outside traffic, and the source and destination of the packets, as well as by any specialized forwarding treatment required for DSCP marked packets.

The top line with arrow shows an example of a UE that is obtaining content from an external source. The content is originally not formulated for the UE's screen, so the OpenFlow controller has installed flow

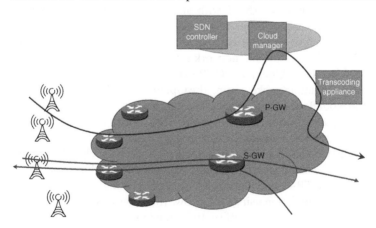

Figure 5.15 Virtualized PC and SDN routing.

rules and actions on the GGSN1, SGSN-D, and the OpenFlow switches to route the flow through a transcoding application in the cloud. The transcoding application reformats the content so that it will fit on the UE's screen. MSC requests the specialized treatment at the time the UE sets up its session with the external content source via the IP multimedia subsystem (IMS) or another signaling protocol.

5.4 Virtualized Customer Premises Equipment

Virtualizing and cloudifying the customer premises equipment (CPE) of enterprises and SMBs is an important NFV use-case. In practice, most existing vCPE PoCs and deployments are overlaid on a distributed physical network topology with relatively static and inefficient resource placement. The current focus has been on connectivity and functionality, rather than performance and agility.

The static resource placement and cumbersome deployment are a result of the need to provide high SLAs on top of resources and tools that were not planned to do so. For example, existing overlay solutions are not aware of the underlay network and its limitations and are hence vulnerable to reduced service levels due to traffic dynamics. In addition, cloud provisioning was designed with compute optimization in mind, whereas VNFs are often bandwidth- rather than compute-constrained. Finally, the industry has not yet come up with NFV-specific SLA monitoring and verification tools that would give customers the assurances and make them trust the lean and dynamic distributed systems with carrying production size loads.

Cloudification of the vCPE solution brings three important advantages:

- *Cost-effective management and agility*: By cloudifying vCPE, we decouple provisioning and VNF onboarding, which is complex and slow today, due to the local provisioning of the physical enterprise site. This semimanual provisioning may be frequent and dependent on many access network factors, including changing demands of the enterprise. In the cloudified solution, the vCPE capacity of all the enterprise sites is in the distributed carrier cloud, for example, the provider point of presence (POP) sites.
- *High performance at scale*: By cloudifying vCPE, we only need to orchestrate resources per POP for the average demand of the

enterprise sites connected to that POP. This includes also Internet peering capacity and related NAT functions. We also do not need to explicitly orchestrate for geo redundant vCPE capacity. This can lead to significant (some estimates are up to 4×) savings in vCPE resources.

- *High availability*: By cloudifying vCPE, we can have faster recovery from vCPE failures. Traffic to failed components is distributed to cloudified resources avoiding sharp hits and potential domino collapses in case of POP or rack failures.

To achieve the goals of cloudification with scalable, dynamic, and efficient operations, we must have the following:

- *Network underlay awareness*: If the demand is randomly distributed to the POP overlay edges, without the knowledge of the underlay, congestion and packet loss could occur, resulting in SLA violations. This happens in the normal course of mapping enterprise flows to vCPE resources and Internet peering, and it is true when remapping enterprise flows in case of failures.
- *SLA verification*: By cloudifying and dynamically mapping traffic to resources, we gain savings and decoupled manageability, but we now have to prove per flow that we still meet the enterprise SLA just as well as with static local provisioning. Per-flow SLA and connectivity verification can also trigger additional VM provisioning without interrupting services.
- *Resource defragmentation*: As we decouple orchestration and the system is running, events will trigger additional allocations, extended service chains, and compensation for blade or CPU failures. But because of cloudification, we can constantly run in the background proceeds that reallocate vCPEs in dense hardware configuration, gracefully ramping down fragmented VNFs till they are "garbage collected" freeing hardware for further dense orchestration.

5.4.1 Requirements

The design of our vCPE platform is based on a set of key requirements we identified by analyzing current SLA structure and management models as well as SMEs common pain points. Our approach highlights elasticity, flexibility, efficiency, scalability, reliability, and openness as critical components to support the goals of NFV.

- *Elasticity*: Building on top of NFV, vCPE should be able to leverage the benefit of running instances in the cloud: multiplexing and dynamical scaling. For multiplexing, it allows the same NF instance to serve multiple end users in order to maximize the resource utilization of the NF. On the other hand, for dynamical scaling, when the demand changes, the network operators should be able to dynamically increase/decrease the number and/or size of each NF type to accommodate the changing demands. This in turn will allow the enterprise customer to benefit from pay-as-you-grow business models and avoid provisioning for peak traffic.

- *Flexibility*: The vCPE platform should support subscriber-based, application-based, device-based, and operator-specific policies simultaneously. Moreover, adding or removing new NFs should be easily manageable by the network operator, without requiring physical presence of technicians on the site or having the enterprise customers involved. It should also be possible to accurately monitor and reroute network traffic as defined by policy. The platform should allow NFs to be implemented, deployed, and managed by operators, enterprises, or third-party software vendors.

- *Efficiency*: The vCPE should provide the tight NF service-level agreements (SLAs) on performance or availability, identical to the SLAs offered with dedicated services. For example, the SLA may specify the average delay, bandwidth, and the availability for all the services provided to one customer. To support the SLA compliance, the platform should closely monitor the performance for each customer and dynamically adapt the resources to meet the SLAs.

- *Scalability*: The vCPE framework should support a large number of rules and scale as the number of subscribers/applications/traffic volume grows. The ability to offer a per-customer selection of NFs could potentially lead to the creation of new offerings and hence new ways for operators to monetize their networks.

- *Reliability*: The vCPE framework should abide by NFV reliability requirements. Service availability as defined by NFV refers to the end-to-end service availability that includes all the elements in the end-to-end service (VNFs and infrastructure components) with the exception of the end-user terminal.

- *Openness*: The final issue is ensuring that the vCPE framework should be capable of accommodating a wide range of NFs in a nonintrusive manner. The vCPE should support open-source-based and standard solutions as much as possible.

5.4.2 Design

In the vCPE architecture, an SME is connected to the Internet via a lightweight CPE also called SDN-based CPE. Most features typically found on a CPE, such as NAT, DHCP, and a firewall/security gateway, are moved to VMs in the operator's cloud. The lightweight SDN-based CPE only retains the basic connectivity service, while being programmable via SDN.

In the operator's cloud, VMs and physical resources (storage, compute) are configured via a cloud controller (CC), while network configuration is managed by an SDN controller. Finally, a moderator node/application provides a higher level of abstraction over the cloud and network controllers. Through the moderator, enterprises can select their desired services and configure service chaining policies. The enterprise's IT personnel can access the moderator through a secured dashboard with a user-friendly graphical interface. Figure 5.16 shows these components and their interactions, which are further described as follows.

- *The customer portal*: Through the customer portal, an enterprise administrator configures and manages enterprise policies, services, and network infrastructure. Each enterprise gets its own virtual infrastructure. VMs are launched for each enterprise and are not shared between enterprises. The first step is to register the vCPE

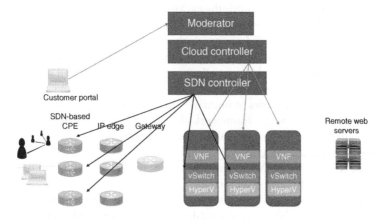

Figure 5.16 Virtualized CPE.

along with its name, IP block, subnets, and so on. Then, through the same portal, the enterprise specifies how the traffic should be mapped and steered across the VNFs (i.e., the service chaining policies).

- The moderator presents services and selected service chains to the enterprise customer, and abstracts away most details of resource allocation for VMs and network configuration. Each enterprise has a catalog of available services. Services can be deployed and chained in arbitrary order in both upstream and downstream directions.

- The CC is a typical cloud controller (e.g., OpenStack) augmented with support for flow networks (i.e., flow network extensions added to the neutron). The moderator translates the list of services and their connectivity into information about VMs, vSwitches, and links that the CC can understand. The CC receives the customer's network architecture and policy specifications, akin to a fine-grained SLA. It translates the SLA into a list of requirements in terms of dedicated VMs, storage, different types of network appliances and business applications, and dedicated links between those. Next, it maps a constructed virtual topology onto the network abstraction provided by the SC (SDN controller). Based on SC feedback, the CC proceeds to create and configure the customer infrastructure (i.e., instantiate VMs, virtual switches). The CC informs the SC of the placement of specific network entities, such as virtual switches.

- The SDN controller (SC) is responsible for managing and provisioning the enterprise network topology, by mapping network requirements to the selected set of physical and virtual network resources (including customer's CPE). Such configuration is done using a combination of different southbound plug-ins, such as OpenFlow, OVSDB, or NetConf. The SC is an SDN controller with a developed application for vCPE service chaining. Some of the extensions require changes to the external interfaces of the SDN controller. The main extensions for vCPE include service chaining, connectivity monitoring, location optimization, and network configuration. Service chaining provided the APIs for orchestrator to create service chaining rules per enterprise. Location optimization and connectivity monitoring provide APIs to detect network congestion and connectivity failures. This information is necessary to enforce network SLAs. Network configuration provides the orchestrator the possibility to engineer service chaining networks in many different ways. The SC will also inform the CC of optimal locations to instantiate and interconnect VMs. It also notifies of link congestion or failure, in order to trigger VM migrations and network reconfiguration.

- The SDN-based CPE and vSwitch is a lightweight version of legacy CPE with most NFs stripped out. The virtual switch is a software-based OpenFlow switch such as Open vSwitch. The VNFs execute in VMs on top of a hypervisor that is hosted on a server. Multiple virtual services can share the resources of the same server. We assume that one instance of a vSwitch is included in each server for handling the communication between multiple VMs running on the servers within the data center. Both SDN-based CPE and vSwitch are programmable by SDN to support the vCPE applications including service chaining.

5.5 Summary

In this chapter, we start with 5G overview. Then, we present the SFC and its several aspects, which is a key use case of SDN and NFV in 5G. We then discuss the SDN and NFV's use cases in the 5G network as two case studies. In particular, we first discuss their usage scenarios and challenges in the packet core network. Next, we discuss their usage in the edge of the telecom network, in the form of customer premise equipment. Both use cases have been widely studied and are going to deployment in the real world.

6

Open Source and Research Activities

Network functions virtualization (NFV) has fundamentally changed the way telecom services are invented and deployed today. In the past, network equipment providers (NEPs) used to sell both the hardware and software. With NFV, the two can be sold separately. Network functions run as software on standardized IT servers. Such separation creates a more open environment for new services to be innovated. It allows communications service providers (CSPs) to be more agile and creative in delivering services, while lowering the cost of development and rollout.

Given the promising features that NFV can provide, CSPs are eager to try out and deploy NFVs in a faster pace. Given this trend, CSPs, NEPs, and NF vendors come together and form open source consortium to expedite this process using a collaborative manner. Open source projects such as OpenStack and OpenDayLight (ODL) are formed. It will encourage greater collaboration among the industry participants. In such open source projects, the basic technologies can be shared across the industry via the upstream projects. Companies can focus on higher-value functions and system integration, which leveraging the nondifferential functions from open source effort from both the CSP and the NEP's perspectives. For the NEPs, they do not need to spend huge time and development resources on the undifferentiated hardware and infrastructure layer, which can be handled by standardized IT and open source software, but instead focusing on their specific network function. The CSPs can leverage the higher-quality code from open source for their infrastructure.

Open source efforts is critical for NFV's growth because it can provide a solid code base for the basic functionalities so that it enables rapid innovation of new services. Traditionally, open source

Network Function Virtualization: Concepts and Applicability in 5G Networks, First Edition. Ying Zhang.
© 2018 John Wiley & Sons, Inc. Published 2018 by John Wiley & Sons, Inc.

has been widely used in the IT environment but not the telecom communication community. Thus, it will be disruptive to every party in the value chain. But on the other hand, it benefits all industry participants as it uses the collective intelligence of the ecosystem to solve the new NFV challenges, which is a critical step toward the NFV transformation.

In this chapter, we introduce various open source initiatives and their focus. For each project, we will start with its goals and scopes, followed by their structure, drivers, and current status.

6.1 Open Source Initiatives

Open source frees the CSPs and NEPs from the development of baseline features that act as the platform upon which everyone can innovate. Many research and survey has shown that CSPs are willing to deploy open source solutions and engage in the open source projects. In the IT industry, there is a long tradition on open source projects, such as Linux and Kernel-based Virtual Machine (KVM). In NFV, several open source projects are developed, each of which has a different focus. They use the same open source model, that is, having a single upstream code base. Contributors include CSPs, NEPs, and NF vendors, each of whom contribute to the common code base in the areas of their specific expertise. Together, they can provide a wide spectrum of the projects. Usually by employing this model, the open source project can leverage a large set of great developers and thus has better quality of code. In this section, we present five different open source activities in the area of Software-defined networking (SDN), NFV, and their usage in telecom networks.

6.1.1 OpenStack

OpenStack is open source cloud computing software that provides infrastructure as a service cloud deployment for public and private cloud [111]. It was started in 2010, when rooted from the NASA's Nebula platform and Rackspace's Cloud Files platform. It is written in python and all of the code for OpenStack is freely available under the Apache 2.0 license. Currently, the projects are managed by OpenStack foundation.

The OpenStack architecture is shown in Figure 6.1. It is organized to cover the three main components in cloud computing: compute,

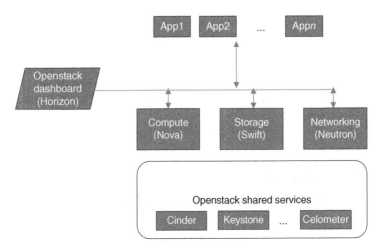

Figure 6.1 OpenStack architecture.

storage, and networking. Besides, dashboard projects are important as they provide administrative interfaces. Their compute component provisions VMs to provide scalable computation resources. The storage component uses objects storage for storing VM images and instances. The networking component provides network services for inter-VM and external network connectivities. Each component and management projects have their own codenames. Examples of some important ones are also shown in Figure 6.1. We highlight some important points of each project as follows:

- Nova is the main core part of OpenStack essential for IaaS. It is responsible for running virtual servers. It is used to host and manage cloud computing systems.
- Swift allows the OpenStack users to store or retrieve files. It stores data in the form of virtual containers, which is based on Rackspace's Cloud Files.
- Neutron is responsible for communication between interface devices, which are managed by other OpenStack services. It provides connectivity as a service to the users.
- Keystone is responsible for handling authentication and authorization in OpenStack. It maintains and manages the rules for users of different levels such as operator, admin, and tenants. Keystone provides authorization token for logging into virtual machines and manages the privileges of users for various services.

- Horizon provides web-based interface to the users to interact with all the OpenStack services, such as managing images, object storage for volumes, and compute for VMs.
- Clinder is the block storage service. It provides a persistent storage for virtual machines in the form of volumes. These volumes are attached with the running instance and act as persistent storage for data storage.
- Celiometer is responsible for various monitoring and metering functionalities.

6.1.2 OpenDayLight

ODL [112] is a collaborative open source project for developing a Java-based open source SDN controller. It is hosted by The Linux Foundation. There are over 50 active members participating in contribution, ranging from NEP, CSP to customers. They can contribute with new initiatives and submit new feature requests to the technical stressing community of ODL.

The ODL architecture is developed based on the Open Services Gateway Imitative (OSGi) framework. Its loosely coupled modular design allows different modules to be constructed independently. Figure 6.2 shows its architectural design. It contains three layers: the top layer is the network applications, the middle layer is the platform controller layer, and the bottom layer is the network elements. The platform controller layer is the essential piece to the controller. It provides the northbound APIs to the applications. This layer is designed modularly. It contains the following three key components:

- The base network service functions gather all the important information and statistics about the network. It includes topology manager that stores the topology, the statistics manager that collects statistical information from managed switches, the forwarding rule manager that handles rule installation and updates, the inventory manager that maintains database of discovered switches, and finally, the host tracker that tracks the location of end hosts in the entire network.
- The platform network service functions perform various networking tasks that serve as service to others. For example, the Affinity metadata service allows application to define certain workloads. The virtual tenant network manager creates and manages multi-tenant virtual network. The L2 switch function provides Layer 2 switching

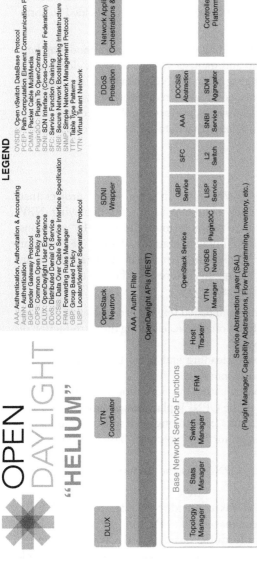

OPEN DAYLIGHT "HELIUM"

LEGEND

AAA: Authentication, Authorization & Accounting
AuthN: Authentication
BGP: Border Gateway Protocol
COPS: Common Open Policy Service
DLUX: OpenDaylight User Experience
DDoS: Distributed Denial Of Service
DOCSIS: Data Over Cable Service Interface Specification
FRM: Forwarding Rules Manager
GBP: Group Based Policy
LISP: Locator/Identifier Separation Protocol

OVSDB: Open vSwitch DataBase Protocol
PCEP: Path Computation Element Communication Protocol
PCMM: Packet Cable MultiMedia
Plugin2OC: Plugin To OpenContrail
SDNi: SDN Interface (Cross-Controller Federation)
SFC: Service Function Chaining
SNBI: Secure Network Bootstrapping Infrastructure
SNMP: Simple Network Management Protocol
TTP: Table Type Patterns
VTN: Virtual Tenant Network

Network Applications Orchestrations & Services

Controller Platform

Southbound Interfaces & Protocol Plugins

Data Place Elements (Virtual Switches, Physical Device Interfaces)

DLUX

VTN Coordinator

OpenStack Neutron

SDNI Wrapper

DDoS Protection

AAA - AuthN Filter

OpenDaylight APIs (REST)

Base Network Service Functions

| Topology Manager | Stats Manager | Switch Manager | FRM | Host Tracker |

OpenStack Service

| VTN Manager | OVSDB Neutron | Plugin2OC |

| GBP Service | SFC | AAA | DOCSIS Abstraction |
| LISP Service | L2 Switch | SNBI Service | SDNI Aggregator |

Service Abstraction Layer (SAL)
(Plugin Manager, Capability Abstractions, Flow Programming, Inventory, etc.)

GBP Renderers

OpenFlow 1.0 1.3 TTP

OVSDB | NETCONF | PCMM/COPS | SNBI | LISP | BGP | PCEP | SNMP | Plugin2OC

OpenFlow Enabled Devices

Open vSwitches

Additional Virtual & Physical Devices

Figure 6.2 OpenDayLight.

functionalities. The previously mentioned service function chaining functionality is also in this category.

- The service abstraction layer enables ODL to support multiple southbound protocols and provide a uniform set of services to other modules. The device discovery service is provided to form the network topology.

Finally, we would like to highlight that ODL uses a modeling language to formally develop and manage various network elements. It uses the yet another next generation (YANG) data modeling language to model network elements, configuration, network states, and so on. It can also be used to model services, protocols, policies, and tenants. It defines data models in modules so that data can be imported/exported between them.

6.1.3 OPNFV

Open platform for NFV (OPNFV) is an open source project that intends to provide an open source platform for deploying NFV solutions [113]. Its goal is to promote interoperable NFV solutions and to create code for NFV implementations. If each NFV vendor implements its own NFV solution, then the interoperability as an industry will be very challenging. More specifically, the goals of OPNFV can be summarized as follows:

- Developing an integrated and tested open source platform that can be used to build NFV functionality, accelerating the introduction of new products and services.
- Including participation of leading end users to validate OPNFV meets the needs of user community.
- Contributing to and participating in relevant open source projects that will be leveraged in the OPNFV platform; ensure consistency, performance, and interoperability among open source components.
- Establishing an ecosystem for NFV solutions based on open standards and software.
- And finally, promoting OPNFV as the preferred open reference platform.

The initial scope of OPNFV is to provide NFV infrastructure (NFVI), virtualized infrastructure management (VIM), and APIs to other NFV elements, which together form the basic infrastructure required for virtualized network functions (VNFs) and management and network orchestration (MANO) components.

The OPNFV project focuses on the NFVI layer: the NFVI and VIM layer. NFVI provides basic access to compute, storage, and network resource from the hypervisor and SDN. The VIM manages VNF images and deploys them in the virtualized environment. Different from the previous open source projects, OPNFV collaborates closely with other open source projects. For VIM, its main upstream project is OpenStack. For its network controller and virtualization infrastructure, its main upstream is ODL. It uses KVM and Xen for the virtualization and hypervisor. Its data plane acceleration uses Data Plane Development Kit (DPDK), another open source project led by Intel. Its first release includes several projects, covering areas of automation, fault tolerance, carrier-grade performance improvements, and network management tools. The details and project outlines are shown in Figure 6.3.

OPNFV works directly with upstream standards bodies such as European Telecommunications Standards Institute (ETSI) and IETF. It also directly works with upstream open source projects, including ODL, OpenStack, KVM and Xen, and many others. It leverages existing codebases so that it can move to integration in a faster pace. It integrates existing open source components and identifies gaps to create new code. Finally, it provides a point of integration, testing, and performance optimization.

Its main organization includes a Business Board and Technical Steering Committee (TSC) governance structure separates business decisions from meritocratic and technical decisions. The board is in charge of nontechnical issues such as auditing, financing, IP and legal, and marketing. The TSC is in charge of projects that drive code development, testing, integration, and reference platform releases. TSC sets technical directions and holds reviews for all the projects.

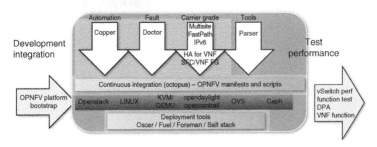

Figure 6.3 OPNFV project overview.

6.1.4   CORD: Central Office Re-architected as a Data Center

Central Office Re-architected as a Data Center (CORD) is an open source project led by the Open Network Lab (ON.Lab) and AT&T [114]. Central offices are of various sizes, supporting a large number of subscribers. Today, central offices have specialized hardware and software to connect the backhaul networks to cellular infrastructure. Typically, central offices contribute to high CAPEX and OPEX, which can be significantly reduced through automation and virtualization. CORD project aims to replace existing Telco central office (CO) infrastructure with data center and cloud components. It contains three key technologies: SDN, NFV, and Cloud. In particular, it would like to use COTS servers and COTS switches as the hardware, which is produced by the Open Compute Project (OCP). From the software side, it uses the ONOS SDN controller. For orchestration, it uses XOS, which is a framework for assembling and composing services. It uses OpenStack to create and manage VMs and virtual networks. Figure 6.4 shows its relationship with other open source projects. ONOS is the network operating system that manages the white box switches. It is a distributed SDN controller. It is also a platform for network services that implement the key CORD functions. Docker or OpenStack is used for virtualization management. The implementation supports both VM and bare metal.

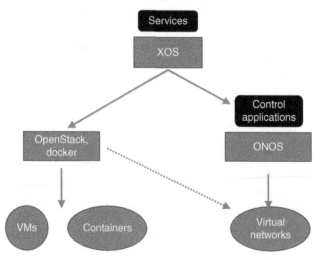

Figure 6.4 CORD architecture.

CORD takes a two-step process to transform the central office. The first step is at the individual device level. It changes the purpose-built hardware devices to commodity hardware where those software can run. This is essentially the NFV concept, which disaggregate and separate software logic from dedicated hardware. The second step in CORD is to provide a framework that can run those software in a scalable manner, using cloud and virtualization technologies.

6.2 NFV Research Problems

In 2014, ETSI published a list of topics in NFV that needs further academic research. The advanced research are needed to bring NFV to the next step for future development. We outline a few of these important areas as follows:

- *Security of the virtualized infrastructure for network functions*: When moving from dedicated integration box to a virtualized environment, the virtualized infrastructure needs to be secured in order to prevent attacks to the upper layer software. While specialized hardware has high capability and thus is harder to be attacked, the virtualization infrastructure may be vulnerable to various exploits and DoS attacks. Making the infrastructure secure requires further research on encryption, trust computing hardware, attack detection, and prevention methods.
- *Abstractions for networks and carrier-scale network services in imperative and declarative languages*: Network programming language has received a lot of attention in the SDN world. In NFV, programmability is rather limited so far. With the microservices and more and more VNFs exposing northbound APIs, the programmability of NFV should be improved via designing flexible and expressive languages.
- *Impacts of data plane workloads on computer systems architectures*: As explained earlier, performance can be improved by exploring shared memory, system on chip (SoC), advanced features in NIC, and so on. Research should be conducted in this area to understand the improvement of NFV from hardware perspective.
- *Locality and latency in software implementations of large-scale network services*: Optimization can be done by exploring the locality of requests. Moreover, various distributed protocols and consistency

mechanisms can be used to support a fully distributed NFV implementation.

- *Re-architecting network functions to recognize availability of cloud technology mechanisms for scalability and reliability*: Some of the network functions, for example, traditional 3GPP and telecom NFs, are not easy to scale since they are not modular. It is important to reconsider its design to fit the new cloud architecture.
- *Evolution patterns to NFV, management of transition and heterogeneous scenarios*: Since the network is going to be evolving gradually, the NFV and the traditional platform will coexist and serve the requests together. Supporting heterogeneous deployment from the management perspective is important for the deployment and evolvement of NFV in the real world.
- *Portability mechanisms and management across NFVI realizations*: There could be multiple virtualization methods, multiple NFVIs, and multiple MANO systems. How to support seamless migration across different platforms is challenging.
- *Tools for validating network services and automating their deployment and management*: Telecom services usually have high requirement on performance and availability, usually requiring five nines of availability. Thus, when deploying services in software, the monitoring and validation is important, especially in face of failure, errors, and human mistakes.
- *Applying compositional patterns (network function chains) for parallelism, control logic, performance, monitoring, and reliability of network services*: Service function chaining can integrate multiple services together. When moving to microservices, that is, each VNF has a unit and simple functionality, merging multiple services together is going to be difficult. Careful design of ordering and parallelism is critical to the performance and the correctness of the entire service chain.

6.3 Summary

In this chapter, we review existing open source activities that are related to SDN, NFV, and can be used in building 5G's new requirements. These projects have wide participants from NEP, CSP, and customers. They are the key enabler for the whole industry and ecosystem.

References

1 AT&T, "AT&T SXSWPress Release,"
2 "Global Mobile Network Traffic–A Summary of Recent Trends," GSMA Documents, 2011, [Online]. http://www.gsma.com/spectrum/wp-content/uploads/2012/03/analysysmasonpaperonglobalmobilenetworktrafficcorectedjuly11.pdf (Accessed November 2016).
3 "Ericsson Mobility Report," November 2012, [Online]. http://www.ericsson.com/res/docs/2012/ericsson-mobility-report-november-2012.pdf (Accessed November 2016).
4 "Ericsson Mobility Report," June 2014, [Online]. http://www.ericsson.com/res/docs/2014/ericsson-mobility-report-june-2014.pdf (Accessed November 2016).
5 "The Zettabyte Era–Trends and Analysis," June 2014. [Online]. http://www.cisco.com/c/en/us/solutions/collateral/serviceprovider/visual-networking-index-vni/VNI_Hyperconnectivity_WP.html (Accessed November 2016).
6 "Cisco Visual Networking Index: Forecast and Methodology, 2013–2018," June 2014. [Online]. http://www.cisco.com/c/en/us/solutions/collateral/service-provider/ip-ngn-ip-next-generationnetwork/white_paper_c11-481360.pdf (Accessed November 2016).
7 R. Sherwood, G. Gibb, K.-K. Yap, G. Appenzeller, M. Casado, N. McKeown, and G. Parulkar, "Can the production network be the testbed?," in *Proceedings of the 9th USENIX Conference on Operating Systems Design and Implementation, OSDI'10*, 2010.
8 N. McKeown, T. Anderson, H. Balakrishnan, G. Parulkar, L. Peterson, J. Rexford, S. Shenker, and J. Turner, "OpenFlow: enabling innovation in campus networks," *ACM SIGCOMM*

Network Function Virtualization: Concepts and Applicability in 5G Networks, First Edition. Ying Zhang.
© 2018 John Wiley & Sons, Inc. Published 2018 by John Wiley & Sons, Inc.

Computer Communication Review, vol. 38, no. 2, pp. 69–74, 2008.

9   V. Sekar, N. Egi, S. Ratnasamy, M. K. Reiter, and G. Shi, "Design and implementation of a consolidated middlebox architecture," in *Proceedings of USENIX NSDI*, 2012.

10 E. N. ISG, "Network Functions Virtualisation (NFV) Architectural Framework," *ETSI GS NFV 002 V1.1.1*, 2013.

11 "VMware Workstation," [Online]. http://www.vmware.com/ca/en/products/workstation (Accessed May 2016).

12 "Network Virtualization Platform," [Online]. http://www.vmware.com/products/nsx.html (Accessed May 2016).

13 [Online]. http://www.openflow.org/.

14 "Microsoft Azure," azure.microsoft.com/.

15 "Google App Engine," cloud.google.com/appengine.

16 "Amazon S3," aws.amazon.com/s3.

17 R. Buyya, C. Vecchiola, and S. T. Selvi, *Mastering Cloud Computing: Foundations and Applications Programming*. Morgan Kaufmann, Elsevier, 2013.

18 K. Ren, C. Wang, Q. Wang, *et al.*, "Security challenges for the public cloud," *IEEE Internet Computing*, vol. 16, no. 1, pp. 69–73, 2012.

19 M. Armbrust, A. Fox, R. Griffith, A. D. Joseph, R. Katz, A. Konwinski, G. Lee, D. Patterson, A. Rabkin, I. Stoica, *et al.*, "A view of cloud computing," *Communications of the ACM*, vol. 53, no. 4, pp. 50–58, 2010.

20 D. Zissis and D. Lekkas, "Addressing cloud computing security issues," *Future Generation Computer Systems*, vol. 28, no. 3, pp. 583–592, 2012.

21 "Docker Container," https://www.docker.com.

22 "runC," https://github.com/opencontainers/runc (Accessed July 19).

23 A. Wang, M. Iyer, R. Dutta, G. N. Rouskas, and I. Baldine, "Network virtualization: technologies, perspectives, and frontiers," *Journal of Lightwave Technology*, vol. 31, no. 4, pp. 523–537, 2013.

24 N. M. K. Chowdhury and R. Boutaba, "Network virtualization: state of the art and research challenges," *IEEE Communications Magazine*, vol. 47, no. 7, pp. 20–26, 2009.

25 N. Chowdhury and R. Boutaba, "A survey of network virtualization," *Computer Networks*, vol. 54, no. 5, pp. 862–876, 2010.

26 S. A. Baset and H. Schulzrinne, "An analysis of the skype peer-to-peer internet telephony protocol," *arXiv preprint cs/0412017*, 2004.

27 E. K. Lua, J. Crowcroft, M. Pias, R. Sharma, S. Lim, *et al.*, "A survey and comparison of peer-to-peer overlay network schemes," *IEEE Communications Surveys and Tutorials*, vol. 7, no. 1–4, pp. 72–93, 2005.

28 P. Knight and C. Lewis, "Layer 2 and 3 virtual private networks: taxonomy, technology, and standardization efforts," *IEEE Communications Magazine*, vol. 42, no. 6, pp. 124–131, 2004.

29 T. Takeda, "Framework and requirements for layer 1 virtual private networks," 2007.

30 Alcatel-Lucent, "VPN Services: Layer 2 or Layer 3?,"

31 R. Sherwood, G. Gibb, K. Yap, G. Appenzeller, M. Casado, N. McKeown, and G. Parulkar, "Flowvisor: a network virtualization layer," *OpenFlow Switch Consortium, Tech. Rep.*, 2009.

32 T. Koponen, K. Amidon, P. Balland, M. Casado, A. Chanda, B. Fulton, I. Ganichev, J. Gross, N. Gude, P. Ingram, E. Jackson, A. Lambeth, R. Lenglet, S.-H. Li, A. Padmanabhan, J. Pettit, B. Pfaff, R. Ramanathan, S. Shenker, A. Shieh, J. Stribling, P. Thakkar, D. Wendlandt, A. Yip, and R. Zhang, "Network virtualization in multi-tenant datacenters," in *Proceedings of the 11th USENIX Conference on Networked Systems Design and Implementation*, NSDI'14, 2014.

33 X. Wang, P. Krishnamurthy, and D. Tipper, "Wireless network virtualization," in *International Conference on Computing, Networking and Communications (ICNC)*, pp. 818–822, IEEE, 2013.

34 H. Wen, P. K. Tiwary, and T. Le-Ngoc, *Wireless Virtualization*. Springer-Verlag, Berlin, 2013.

35 L. Doyle, J. Kibiłda, T. K. Forde, and L. DaSilva, "Spectrum without bounds, networks without borders," *Proceedings of the IEEE*, vol. 102, no. 3, pp. 351–365, 2014.

36 R. Chandra and P. Bahl, "MultiNet: Connecting to multiple IEEE 802.11 networks using a single wireless card," in *Proceedings of IEEE INFOCOM*, vol. 2, pp. 882–893, 2004.

37 Microsoft Research, "Connecting to Multiple IEEE 802.11 Networks with One WiFi Card," [Online]. http://research.microsoft.com/en-us/um/redmond/projects/virtualwifi/ (Accessed August 2016).

38 L. Xia, S. Kumar, X. Yang, P. Gopalakrishnan, Y. Liu, S. Schoenberg, and X. Guo, "Virtual WiFi: bring virtualization from wired to wireless," in *ACM SIGPLAN Notices*, vol. 46, pp. 181–192, ACM, 2011.

39 Y. Al-Hazmi and H. de Meer, "Virtualization of 802.11 interfaces for wireless mesh networks," in *Proceedings of the 8th International Conference on Wireless On-demand Network Systems and Services (WONS)*, pp. 44–51, 2011.

40 F. Boccardi, O. Aydin, U. Doetsch, T. Fahldieck, and H. Mayer, "User-centric architectures: enabling CoMP via hardware virtualization," in *Proceedings of IEEE International Symposium on Personal Indoor and Mobile Radio Communications (PIMRC)*, pp. 191–196, 2012.

41 D. Lee, H. Seo, B. Clerckx, E. Hardouin, D. Mazzarese, S. Nagata, and K. Sayana, "Coordinated multipoint transmission and reception in LTE-advanced: deployment scenarios and operational challenges," *IEEE Communications Magazine*, vol. 50, no. 2, pp. 148–155, 2012.

42 G. Smith, A. Chaturvedi, A. Mishra, and S. Banerjee, "Wireless virtualization on commodity 802.11 hardware," in *Proceedings the 2nd ACM International Workshop on Wireless Network Testbeds, Experimental Evaluation and Characterization*, pp. 75–82, ACM, 2007.

43 A. Gudipati, D. Perry, L. E. Li, and S. Katti, "SoftRAN: Software Defined Radio Access Network," in *Proceedings the 2nd ACM SIGCOMM Workshop on Hot Topics in Software Defined Networking*, pp. 25–30, ACM, 2013.

44 S. Katti and L. E. Li, "RadioVisor: A Slicing Plane for Radio Access Networks," in *Presented as Part of the Open Networking Summit 2014 (ONS 2014)*, (Santa Clara, CA), USENIX, 2014.

45 K. Nakauchi, K. Ishizu, H. Murakami, A. Nakao, and H. Harada, "AMPHIBIA: a cognitive virtualization platform for end-to-end slicing," in *Proceedings of IEEE International Conference on Communications (ICC)*, pp. 1–5, IEEE, 2011.

46 Z. Zhu, P. Gupta, Q. Wang, S. Kalyanaraman, Y. Lin, H. Franke, and S. Sarangi, "Virtual base station pool: towards a wireless network cloud for radio access networks," in *Proceedings of the 8th ACM International Conference on Computing Frontiers*, p. 34, ACM, 2011.

47 J. G. Andrews, S. Buzzi, W. Choi, S. Hanly, A. Lozano, A. C. Soong, and J. C. Zhang, "What will 5g be?," *arXiv preprint arXiv:1405.2957*, 2014.

48 S. Singhal, G. Hadjichristofi, I. Seskar, and D. Raychaudhri, "Evaluation of UML based wireless network virtualization," in *Proceedings of the Next Generation Internet Networks*, 2008.

49 M. Pearce, S. Zeadally, and R. Hunt, "Virtualization: issues, security threats, and solutions," *ACM Computing Surveys*, vol. 45, no. 2, p. 17, 2013.

50 "Network Functions Virtualisation," [Online]. http://www.etsi.org/technologies-clusters/technologies/nfv (Accessed November 2016).

51 L. Cao, P. Sharma, S. Fahmy, and V. Saxena, "NFV-VITAL: a framework for characterizing the performance of virtual network functions," in *IEEE SDN-NFV Conference*, 2015.

52 A. Gember-Jacobson, R. Viswanathan, C. Prakash, R. Grandl, J. Khalid, S. Das, and A. Akella, "OpenNF: Enabling innovation in network function control," in *Proceedings of the 2014 ACM Conference on SIGCOMM*, SIGCOMM '14, 2014.

53 J. Khalid, A. Gember-Jacobson, R. Michael, A. Abhashkumar, and A. Akella, "Paving the way for NFV: simplifying middlebox modifications using statealyzr," in *13th USENIX Symposium on Networked Systems Design and Implementation (NSDI 16)*, pp. 239–253, 2016.

54 S. K. Fayaz, T. Yu, Y. Tobioka, S. Chaki, and V. Sekar, "Buzz: Testing context-dependent policies in stateful networks," in *Proceedings of USENIX NSDI*, 2016.

55 B. Tschaen, Y. Zhang, T. Benson, S. Benerjee, J. Lee, and J.-M. Kang, "SFC-checker: checking the correct forwarding behavior of service function chaining," in *IEEE SDN-NFV Conference*, 2016.

56 R. Stoenescu, M. Popovici, L. Negreanu, and C. Raiciu, "Symnet: Scalable symbolic execution for modern networks," in *Proceedings of ACM SIGCOMM*, 2016.

57 W. Wu, Y. Zhang, and S. Banerjee, "Automatic synthesis of NF models by program analysis," in *Proceedings of the 15th ACM Workshop on Hot Topics in Networks*, HotNets '16, pp. 29–35, 2016.

58 M. Weiser, "Program slicing," in *Proceedings of the 5th International Conference on Software Engineering*, pp. 439–449, IEEE Press, 1981.

59 H. Agrawal and J. R. Horgan, "Dynamic program slicing," in *ACM SIGPLAN Notices*, vol. 25, pp. 246–256, ACM, 1990.

60 S. Horwitz, T. Reps, and D. Binkley, "Interprocedural slicing using dependence graphs," *ACM Transactions on Programming Languages and Systems (TOPLAS)*, vol. 12, no. 1, pp. 26–60, 1990.

61 J. Ferrante, K. J. Ottenstein, and J. D. Warren, "The program dependence graph and its use in optimization," *ACM Transactions on Programming Languages and Systems (TOPLAS)*, vol. 9, no. 3, pp. 319–349, 1987.

62 V. Chipounov, V. Kuznetsov, and G. Candea, "S2E: a platform for in-vivo multi-path analysis of software systems," *ACM SIGPLAN Notices*, vol. 46, no. 3, pp. 265–278, 2011.

63 C. Cadar, D. Dunbar, D. R. Engler, *et al.*, "KLEE: unassisted and automatic generation of high-coverage tests for complex systems programs," in *OSDI*, vol. 8, pp. 209–224, 2008.

64 M. Dobrescu and K. Argyraki, "Software dataplane verification," in *11th USENIX Symposium on Networked Systems Design and Implementation (NSDI 14)*, pp. 101–114, 2014.

65 D. Joseph and I. Stoica, "Modeling middleboxes," *IEEE Network - The Magazine of Global Internetworking*, 2008.

66 A. Panda, K. Argyraki, M. Sagiv, M. Schapira, and S. Shenker, "New directions for network verification," in *LIPIcs-Leibniz International Proceedings in Informatics*, vol. 32, Schloss Dagstuhl-Leibniz-Zentrum fuer Informatik, 2015.

67 A. Panda, O. Lahav, K. Argyraki, M. Sagiv, and S. Shenker, "Verifying reachability in networks with mutable datapaths," in *Proceedings of USENIX NSDI*, 2017.

68 S. Zhu, J. Bi, C. Sun, and C. Wu, "SDPA: Enhancing stateful forwarding for software-defined networking," in *Proceedings of IEEE ICNP*, 2015.

69 M. Moshref, A. Bhargava, A. Gupta, M. Yu, and R. Govindan, "Flow-level state transition as a new switch primitive for SDN," in *Proceedings of the 2014 ACM Conference on SIGCOMM*, SIGCOMM '14, 2014.

70 G. Bianchi, M. Bonola, A. Capone, and C. Cascone, "OpenState: programming platform-independent stateful openFlow applications inside the switch," *SIGCOMM Computer Communication Review*, vol. 44, pp. 44–51, 2014.

71 "http://www.haproxy.org/,"

72 "http://www.gedanken.org.uk/software/wwwoffle/,"

73 D. Angluin, "Learning regular sets from queries and counterexamples," *Inf. Comput.*, vol. 75, pp. 87–106, 1987.

74 P. Kazemian, G. Varghese, and N. McKeown, "Header space analysis: static checking for networks," in *Proceedings of USENIX NSDI*, 2012.

75 H. Mai, A. Khurshid, R. Agarwal, M. Caesar, P. Godfrey, and S. T. King, "Debugging the data plane with anteater," *ACM SIGCOMM Computer Communication Review*, vol. 41, pp. 290–301, 2011.

76 "https://www.inlab.de/balance.html,"

77 "Snort IDS Web page," http://www.snort.org/.

78 M. Forzati, C. Larsen, and C. Mattsson, "Open access networks, the Swedish experience," in *Proceedings of the 12th IEEE International Conference on Transparent Optical Networks*, pp. 1–4, 2010.

79 S. Jain, A. Kumar, S. Mandal, J. Ong, L. Poutievski, A. Singh, S. Venkata, J. Wanderer, J. Zhou, M. Zhu, J. Zolla, U. Hölzle, S. Stuart, and A. Vahdat, "B4: experience with a globally-deployed software defined WAN," *SIGCOMM Computer Communication Review.*, vol. 43, no. 4, pp. 3–14,2013.

80 T. Koponen, M. Casado, N. Gude, J. Stribling, L. Poutievski, M. Zhu, R. Ramanathan, Y. Iwata, H. Inoue, T. Hama, and S. Shenker, "Onix: a distributed control platform for large-scale production networks," in *Proceedings of the 9th USENIX Conference on Operating Systems Design and Implementation*, OSDI'10, 2010.

81 "http://onosproject.org/,"

82 P. Sun, R. Mahajan, J. Rexford, L. Yuan, M. Zhang, and A. Arefin, "A network-state management service," *SIGCOMM Computer Communication Review*, vol. 44, pp. 563–574, 2014.

83 "Network Intent Composition," https://wiki.opendaylight.org/view/Network_Intent_Composition:Main.

84 A. AuYoung, Y. Ma, S. Banerjee, J. Lee, P. Sharma, Y. Turner, C. Liang, and J. C. Mogul, "Democratic resolution of resource conflicts between sdn control programs," in *Proceedings of the 10th ACM International on Conference on Emerging Networking Experiments and Technologies*, CoNEXT '14, 2014.

85 T. Zseby, "Netflow,"

86 Y. Zhang, "An adaptive flow counting method for anomaly detection in SDN," in *Proceedings of the 9th ACM Conference on Emerging Networking Experiments and Technologies*, CoNEXT '13, 2013.

87 X. Liu, M. Shirazipour, M. Yu, and Y. Zhang, "MOZART: temporal coordination of measurement," in *Proceedings of the Symposium on SDN Research*, SOSR '16, pp. 13:1–13:12, 2016.

88 http://www.internet2.edu/network/.

89 Y. Zhang, N. Beheshti, and M. Tatipamula, "On resilience of split-architecture networks," *In Proceedings of IEEE GLOBECOM 2011 - Next Generation Networking Symposium*, 2011.

90 The New York Times, "How the Cyberattack on Spamhaus Unfolded," http://www.nytimes.com/interactive/2013/03/30/technology/how-the-cyberattack-on-spamhaus-unfolded.html.

91 "The new threat: Targeted internet traffic misdirection," http://research.dyn.com/2013/11/mitm-internet-hijacking/, 2014.

92 A. Keromytis, V. Misra, and D. Rubenstein, "SOS: an architecture for mitigating DDoS attacks," *IEEE Journal on Selected Areas in Communications*, vol. 22, no. 1, pp. 176–188, 2004.

93 M. T. Goodrich, "Probabilistic packet marking for large-scale IP traceback," *IEEE/ACM Transactions on Networking*, vol. 16, pp. 15–24, 2008.

94 J. Ioannidis and S. M. Bellovin, "Pushback: Router-based defense against ddos attacks," in *NDSS*, 2002.

95 G. Oikonomou, J. Mirkovic, P. Reiher, and M. Robinson, "A framework for a collaborative ddos defense," in *ACSAC '06: Proceedings of the 22nd Annual Computer Security Applications Conference*, pp. 33–42, IEEE Computer Society, 2006.

96 X. Liu, X. Yang, D. Wetherall, and T. Anderson, "Efficient and secure source authentication with packet passports," in USENIX SRUTI, 2006.

97 S. Jain, A. Kumar, S. Mandal, and J. Ong, "B4: experience with a globally-deployed software defined WAN," in *Proceedings of ACM SIGCOMM*, 2013.

98 A. Kumar *et al.*, "BwE: flexible, hierarchical bandwidth allocation for WAN distributed computing," in *Proceedings of ACM SIGCOMM*, 2015.

99 C.-Y. Hong *et al.*, "Achieving high utilization with software-driven WAN," in *Proceedings of ACM SIGCOMM*, 2013.

100 N. Megiddo, *On the Complexity of Linear Programming*. IBM Thomas J. Watson Research Division, 1986.

101 T. A. Forum *et al.*, "Private Network-Network Interface Specification Version 1.0 (PNNI 1.0)," 1996.

102 I. Castineyra *et al.*, "The Nimrod Routing Architecture," in *IETF, RFC*, 1992.

103 M. Moradi *et al.*, "SoftMoW: recursive and reconfigurable cellular WAN architecture," in *Proceedings of ACM CoNEXT*, 2014.

104 H. H. Liu *et al.*, "Traffic engineering with forward fault correction," in *Proceedings of ACM SIGCOMM*, 2014.

105 L. Fang *et al.*, "Hierarchical SDN for the hyper-scale, hyper-elastic data center and cloud," in *Proceedings of ACM SOSR*, 2015.

106 Z. A. Qazi, C.-C. Tu, L. Chiang, R. Miao, V. Sekar, and M. Yu, "Simple-fying middlebox policy enforcement using SDN," in *Proceedings of the ACM SIGCOMM*, SIGCOMM '13, pp. 27–38, 2013.

107 S. K. Fayazbakhsh, L. Chiang, V. Sekar, M. Yu, and J. C. Mogul, "Enforcing network-wide policies in the presence of dynamic middlebox actions using flowtags," in *Proceedings of the USENIX Conference on Networked Systems Design and Implementation*, pp. 533–546, 2014.

108 Y. Zhang, N. Beheshti, L. Beliveau, G. Lefebvre, R. Manghirmalani, R. Mishra, R. Patneyt, M. Shirazipour, R. Subrahmaniam, C. Truchan, and M. Tatipamula, "StEERING: a software-defined networking for inline service chaining," in *Proceedings of IEEE International Conference on Network Protocols*, pp. 1–10, October 2013.

109 "OpenFlow 1.1." http://www.openflow.org/wk/index.php/OpenFlow_v1.1.

110 A. Khurshid, X. Zou, W. Zhou, M. Caesar, and P. B. Godfrey, "VeriFlow: verifying network-wide invariants in real time," in *Proceedings of USENIX NSDI*, 2013.

111 "Openstack," https://www.openstack.org/.

112 "Opendaylight," https://www.opendaylight.org.

113 "Opnfv," https://www.opnfv.org/.

114 "Cord," https://www.opencord.org/.

Index

Network Function Virtualization: Concepts and Applicability in 5G Networks, First Edition. Ying Zhang.
© 2018 John Wiley & Sons, Inc. Published 2018 by John Wiley & Sons, Inc.

Printed and bound by CPI Group (UK) Ltd, Croydon, CR0 4YY

23/04/2025

14660906-0001